There's a calf in the sitting room...

A cow face to face

Has charm and grace

But if exploitation's

Your preoccupation

You'll have to befriend

The other end.

There's a calf in the sitting room...

Sheila Barry

breedon **books**
PUBLISHING

First published in Great Britain in 2003, this edition 2006 by
The Breedon Books Publishing Company Limited
Breedon House, 3 The Parker Centre,
Derby, DE21 4SZ.

Dedication

In memory of
Madge and Cecil and Llangrannog.

ISBN 1 85983 529 5

Printed and bound by Biddles Ltd, King's Lynn, Norfolk, UK

Contents

Preface 7

Acknowledgements 9

Foreword 10

Prologue 11

Chapter 1   Beginnings 19

Chapter 2   Language and Sayings 24

Chapter 3   The Amenities 31

Chapter 4   More Family Members 37

Chapter 5   There Goes the Hooter 47

Chapter 6   No Snow to Speak of 56

Chapter 7   Surviving 63

Chapter 8   Pig Killing 70

Chapter 9   Sleeping Out 78

Chapter 10   Everyday Jobs 85

Chapter 11   A Very Special Year 94

Chapter 12   Not Suburbia 102

Chapter 13   If Mrs Jones Can Do It, I Can Do It 109

Chapter 14   Crops 118

Chapter 15   Paying Guests 128

Chapter 16   Haymaking 135

Chapter 17   Our Modernest Conveniences 144

Chapter 18   Cliff Hangers 152

Chapter 19   Corn Harvest 160

Chapter 20   Cardi Connections 170

Chapter 21   Party Pieces 177

Annibendod 184

Index 190

Penrallt 1960. An Atcost shed in haggard replaces the old dutch barns. The pond is at the far end of the hedged oval behind the cowshed. Camping huts and Jimmy Burton's caravan are by Home Field hedge. The lanes run roughly NE by SW. The milkstand road, or crossing lane, runs past the house at top left (Morwel and Bryneuron in trees) and on through Llainwen crossroads at the buildings in the top right of the picture. Clumps of trees planted by Cranogwen may be seen in the corners of the top six fields below the road.

Looking across the farm to Llangrannog. The road to Pontgarreg runs round the base of Lochtyn. The first cluster of houses, 200ft above the sea and hidden from it by the hill, is where the original village and church enclosure were built, around AD 500.

Preface

'What a clever title.'

'How did you think of such a good title?'

'I saw the title and stopped to have another look.'

Comments like these tended to come quite often from readers talking to me about my book. I say 'tended', past tense, because after its very successful and complete sell-out, I haven't met many new readers lately!

In fact, when I first started putting this book together, I was calling it *If Mrs Jones Can Do It, I Can Do It*. This was in tribute to the retiring farmer's wife, a little old lady half Madge's size, who told my aunt Madge how to deal with the woman's side of farming. Most of this advice and instruction Madge followed when she could, but there was one thing she found quite unacceptable. This was the custom of the men sitting down and being waited on by the women, who only sat and ate themselves when the men had all finished and gone. Madge organised dinners for everyone to sit and eat together; except when there were two sittings of a big working gang, as with threshing. There was also one notorious occasion when she was completely unable to do what Mrs Jones could do, despite repeating the mantra over and over as she tried. This was her confrontation with Mrs Goose, described later in the chapter entitled 'If Mrs Jones Can Do It, I Can Do It'. This was when the saying passed into the Collins book of proverbs.

However, as Susan Last, my editor at Breedon Books, pointed out, this title told prospective readers nothing about the content of the book, nothing of the essential character and interest of the story. Something more descriptive was required. Something that would grab the casual browser and make her say, 'That looks interesting'; and so inspire her to look again more closely. Between us we came up with *There's a Calf in the Sitting Room*, which did indeed prove a great attraction to buyers, as I have said. Even more eye-catching was the clever choice of pictures and their arrangement on the cover. Many people commented on this, and the jolly little pictures put with the chapter headings. They said what a beautifully made book it was.

Older people were delighted to have the old ways written about. Many others, particularly younger and non-farming people, said how interesting it was to see into the complexities of the early machines, which they had previously dismissed as simple, primitive constructions. They were astonished, too, at the skills required to manoeuvre, work and maintain them.

All of this appreciation was particularly pleasing as my intention when starting to write had been to keep alive a knowledge of the old ways of farming before modern machinery changed the entire concept of agriculture and before labour-saving amenities changed people's expectations and attitudes to life. Although it is set in a particular time and place, West Wales at the end of World War Two, farming customs and conditions of life were much the same in other remote parts of the British Isles. People in East Anglia, Oxfordshire, Lancashire and Jersey, as well as various parts of Wales, remembered the threshing machines and their gangs of workers going from farm to farm. One man in Oxfordshire talked of helping to build corn stacks with his father when he was five years old. He learnt to tuck the sheaves of corn in tight together, layering them to make a safe tidy stack that would stand firm and dry through the winter. I had letters from people who were not farmers but lived in villages or on the very fringes of towns. Here too, until well into the 1950s, there was no electricity or running water. Light came from paraffin lamps, water had to be carried from a well and the toilet was an earth closet down the garden. Several

Griffy Jones, Penrallt, 1946.

Madge Collins, Penrallt, 1960.

ladies remembered blackleading the old grates and I had letters saying, we had to go outside to the privy; we had candles and paraffin lamps; I had to pump and carry water.

All of these things, I thought, needed to be written down. Not just the bare outlines, but details of how the machines worked and the skills needed for handling them, for these were much more complicated than you might expect. Even lighting a paraffin lamp was not a simple matter. In fact there was an innate malevolence in all paraffin gadgets. It waited to leap out like a bad genie if you failed to perform the magic rituals correctly.

But the technical details on their own were not enough. They needed to be described accurately for experts, but also as simply as possible for those only marginally interested. And to hold it all together as an interesting, readable book it needed stories of the farmers themselves. Here I was lucky in having not only the everyday farming life to describe, but the added spice of a rather eccentric family and their friends. This mix seemed to work out well, for all sorts of people said how much they'd enjoyed the book, even though they skimmed over the technical bits. Others found my descriptions of the machinery and its use added to their enjoyment and understanding.

Although I mention many local people, my intention was not to fill the book with the usual run of country anecdotes. The one scandal I do tell, really because it shows the sadness and humour of such things and how they can spread into lasting feuds and folklore, I keep deliberately vague. Even so I heard that one local lady complained and said that I should not have written about this. She was so outraged she tore out the offending pages. I told this to another of us few remaining Early Inhabitants of Llangrannog, pointing out with some indignation that I had taken great care to name no names, not even of the villages concerned. No-one need have been offended.

'Ah,' she said slyly, 'but perhaps she was one of the names!'

Although farming methods and machinery were already changing by 1953, these changes did not really alter the way of life noticeably. Farming was still very much a hands-on, open-air business, with farmers walking their fields and talking to their cows, handling their crops. Practically every farm now had at least one tractor, many of them with cabs; though these only kept out the very worst of the rain and wind and you still had to get out and do a lot of heavy lifting and shoving to fix the implements. Not like the modern-day tractors where you sit in a miniature hotel room with armchair, radio and central heating and just push a few knobs and levers to pick things up and put them down.

I touch on some of these coming changes in *Theres's A Calf in the Sitting Room* and I am hoping to write a sequel, which will cover the next stage in farming life and my experience of it. I hope it will turn out well enough for Breedon Books to want to publish it and for you lot to want to read it.

Sheila Barry
April 2006

Some readers' comments:

'Informative, vibrant and refreshingly different.'
'I wanted to thank you for writing down all those important and nostalgic memories in your delightful book.'
'Beautifully laid out. A pleasure to read.'
'An authentic glimpse of the past.'
'Interesting and informative. I had no idea what was involved working on a farm.'

Foreword

This book is set in West Wales just after World War Two, when food, and virtually all other commodities you can think of, were scarce, rationed, or non-existent. It was a time when farming was almost organic and mostly horse-driven. Towns were connected to the national grid electricity supply, but villages and isolated houses were not. Nor was there universal piped water.

This book is about how farming was done in that part of the country at that time, and how it began to change. And it is about the people of the area and how they lived and changed. In particular it is about one rather eccentric family of 'In-Comers' and how they and the local community reacted together.

Acknowledgements

My thanks to my husband, Patrick, for diagrams and all technical information. Without him, this book would have been less interesting and informative.

Prologue

THREE of us stood in the cold little room.

The man in charge drew back the cover and I opened my mouth to say: Yes. That's my mother.

But it wasn't.

'No.'

I could hear the surprise in my voice. I had come to this identification of the body in a matter-of-fact, level-headed way. I knew it had to be my mother but look as I might I could see no sign of her. This was a wax dummy.

The others were waiting.

'No,' I repeated. 'No it's not.' What else could I say?

The man in charge of the viewing looked enquiringly at my friend, a mature young woman of 22 years to my very immature 17. I saw her nod. The cloth was replaced and we left. The chauffeur-driven car kindly loaned for the afternoon, along with my friend, Pat, by the Office Manager of the Wages Department of Cranes Ltd, Ipswich, took us back to work.

It was Monday 11 November, Armistice Day. Dry and sunny, I seem to remember. World War Two had ended more than a year ago, but though the horror and misery of the killing was over, austerity in everyday life remained. A grey shabbiness pervaded everything, from make-do-and-mend clothes to pale, glimmering street lights. Virtually all food was rationed and many everyday things were scarce or unobtainable. Bananas and oranges, unknown to children born during the war, were just beginning to arrive in grocers' shops. Clothes, soap and coal were all rationed. And sweets. Gas and electricity supplies were restricted so that when demand was too high, everybody's lights dimmed, oven heat diminished and other gadgets faded or gave up working altogether. There was a very small allocation of petrol for businesses, for which another book of coupons was issued. Being shopkeepers we had a petrol ration, and out of this we managed to make a few non-business drives into the country or to relatives.

It was during one of these drives that the accident occurred. Dad had recently been diagnosed as having what was then called disseminated sclerosis, now multiple sclerosis, and he was starting to teach my mother to drive. On Sunday 10 November 1946, another fine day, they set out for the quiet country roads north of Ipswich.

As they were leaving, Dad was joking about the possibility of Mum crashing the car. 'You'd better notify the police if we're late.'

'What do you call late?'

'Oh, around tea-time. We shouldn't be later than five.'

As we were speaking, a subtle change had crept into the atmosphere. We were no longer joking. This is not something I have imagined with hindsight. A chilling apprehension hung in the air. I knew Dad felt it too. I was not aware of my mother. She must have left the room. We were not a superstitious or intuitive family. Only Mum had leanings towards fortune tellers and friends who were spiritualists. So of course it was just an intrusion of seriousness into the conversation, knowing the possibility of an accident was always there. If it was a warning or other form of premonition, we chose, I suppose, to ignore it.

'We'll be back before dark,' my father said then, matter-of-factly.

I don't know what time the policeman came but it had been dark for a long time. I have in mind seven o' clock, but it was possibly earlier. I had gone through the likely scenario: my father was dead, mercifully released from the prognosticated year's slow, painful dying, and Mum and I would get a cottage in the country with a pig, a donkey and some chickens. It was something we had talked half-seriously about sometimes, though it was probably more my dream than hers. It was impossible anyway as the shop would have had to go on providing a living.

When the knock came, I opened the door and the policeman entered the dark hallway. Wartime economy persisted and the electricity supply was low. He prowled along the passageway while I sat on the stairs and watched him through the banisters. I don't know if he spoke first or if I asked what had happened.

There had been an accident, he said. My father was badly injured and in hospital.

'My mother's all right then? Is she coming home?' He hadn't mentioned her. She must have escaped unharmed.

'I'm afraid she won't be coming home.'

'Oh. Why not? You didn't say she was in hospital.'

'She's not in hospital.'

'Why isn't she coming home then?' Then I realised. Of course! She'd be with Dad.

'She isn't coming home because she's dead.'

'No.' I was bewildered. 'Oh. – No.'

I didn't believe it. That my bright, full of life mother would be dead had never once occurred to me. It wasn't possible. I peered at the poor man through the bars, trying to

Galloping Major, E27N, pulling old Morris out to get him started.

rearrange my story. It must have been a very unnerving experience for him. He must have dreaded the job anyway. Would it have been easier or worse for him if I had screamed and cried? We talked of getting in touch with relatives, of my being all right on my own, of visiting my father. When he left, I got on my bike and cycled the two miles or so to the hospital. My father lay in a coma. Thankfully he died four days later without regaining consciousness.

I was on my own in the flat, my older brother away in the Army of Occupation in Germany. We had no telephone, nor had our close relations or anyone I knew, so the news was passed on by letter. Two days after the accident my Auntie Kathleen, Mum's sister, arrived to sort things out. She spent a lot of time sitting with my father and was with him when he died.

I remember little of the funeral. It was drab and remote. A cold, grey day but dry. A very disparate group of people packed into the tiny flat above our shop. I'm not even sure who they all were. Isolated moments stick in my mind like photos on a shelf: a mass of flowers; sitting in a car beside my brother and being driven to the cemetery; standing beside the earthy hole; the coffins being lowered. Why were there two? Why couldn't they have shared one?

The packed room again, this time rowdy with eating and drinking. Who arranged that? Auntie Kathleen, I suppose. My brother, home on compassionate leave from the army in Germany, was drunk and kept reeling against the sideboard, causing a delicate, engraved brass tray which lodged precariously on the upper shelf to keep falling down. It finally cracked and was set aside.

Among the mourners were my father's brother, Cecil, and his wife Madge. I had met this aunt and uncle for the first time, very briefly, a few months earlier. They were our family 'black sheep' and rarely mentioned, at least not in my hearing. Personal information about relations, or any adult, was no business of the children and consequently I know virtually no family history. There were people who would have told me some of it – my grandmother, Auntie Kathleen, – but I never asked because I was always told not to ask questions, not to stare, not to listen to other people's conversations, and I always tried to do as I was told because life was slightly easier that way. One piece of information that had seeped through in some lax moment was that Cecil had left his wife and small son and gone off with Another Woman. The only other mention of him came at times when my brother was behaving in ways our parents found utterly unacceptable and beyond their understanding, when it was said he must take after Cecil.

These remote and intriguing people were just in the process of buying a farm in Wales. To my delight, they invited me to go and stay. Living on a farm had always been one of my dreams. Part of my 'What I Would Like to do When I Grow Up' compositions. The other part was writing.

Auntie Kathleen was expecting me to go and live with her, but much as I loved her, I could not face living in a family that would be the affectionate, stifling opposite to the rather detached, aggressive attitude I was used to from my mother.

I said I intended to stay living and working in Ipswich and, slightly to my surprise, I got away with this without great argument. When the shop was sold I moved into the YWCA, which eased Auntie Kate's worries slightly. Despite sadness for my parents, I really enjoyed the weeks of independence that followed.

At this time I had been working for just over a year in the wages department of Cranes Ltd, a factory that made boilers, I think, and associated bits and pieces. In that time my wages had risen from 16s 9d a week to £1 and a few pence. Almost immediately after my parents' death, this was raised to £3 3s 0d because I now had to support myself. Things were done differently in those days.

The violent slamming of one door opened two or three others for me, and of them I took neither the most nor the least adventurous. I chose, as I have done all my life, the compromise. Unlike the straight firm road of decision, compromise meanders through fords and forests and along enticing, unsignposted byways.

My life has been a fairly indecisive meander.

Even going to Wales was a fanciful event. It may sound like a decision, but it was no more than a dream really. Here was this aunt and uncle I didn't know saying I might go and stay any time I liked. Stay as long as I liked, they said. And though from that first invitation I knew in my guts that I was going to go and I was going to stay for ever, I didn't say this in so many words, even to myself. I hedged the thought around, against disappointment, with dismissive shrugs and ambiguous smiles. But somehow I did know.

When the invitation was repeated on a Christmas card, I began to think, why wait till summer? I kept up the appearance of going for no more than a short holiday by taking only a small suitcase. The only thing I possessed remotely suitable for farming was the zip jacket I wore for cycling. I went to the new, backstreet shop that sold ex-service personnel clothing and bought a pair of WAAF trousers. These were of rough serge material about half an inch thick. I lived in them for three months, though I did have a skirt to change into.

I cancelled my room at the YWCA, but not my job. I think I never did do that. I was giving no one the chance to say 'I told you so'.

Snow fell in Ipswich that Christmas, but on the opposite coast in Llangrannog, Wales, almost a straight line from east to west, the weather, as I learned later, was like spring. People I came to know intimately were walking and paddling on a beach which had never known the restrictions of concrete and barbed wire, and had never been a forbidden area as our east coasts had been and still were.

It was 20 January 1947 when I set off.

I wore my mother's mid-blue astrakhan coat, which reached below calf length on me, and her pill-box hat of blue feathers with deep red feathers on top. I saw her wear this hat only once or twice and I suspect she fell in love with it as I did, but had few occasions to dress up and wear it. I thought these clothes were gorgeous and in them I felt assured and sophisticated. Trying now to visualise myself, I think I might even have looked quite exotic and not utterly ridiculous.

Trains were slower in those days and always packed so full you considered yourself lucky to find space in the corridor to put your bag down and sit on it. Moving about involved clambering over legs and luggage and squeezing intimately past people. Most of those travelling were servicemen, and a deal of cheery flirting and suggestive banter went on, the sort of thing that today constitutes sexual harassment and politically incorrect behaviour. Then it was part of the fun and excitement of life. Offence was rarely given or taken. Life was immensely different in those days in a society where sex was not acknowledged in public and still retained the excitement of the unknown and forbidden.

The journey was cold and tiring and there was an indecisiveness about it. I thought when we entered Wales that I must be almost at my destination, but counting in hours rather than miles, this was little more than halfway.

When I got off at Carmarthen, I was greeted by the Station Master with the news that my uncle had been unable to get his car started and there was a taxi to take me the rest of the way.

The long, winding 30 miles from Carmarthen to Llangrannog seemed to go on for ever. It probably took us a good hour and a half and in all that time and distance I believe we saw no other traffic at all.

It must have been cold but I don't remember particularly. Cars were draughty, rattly things with no heaters, but one was used to being cold. I do remember the noise though: rain beating down, water swishing up from holes in the road as we rattled through them. And all accompanied by sudden timpani trills as the wind whipped rain and branches against our sides.

I remember the darkness too. The wartime blackout I had recently lived through had never been so devoid of light as this land.

Our low-grade headlights showed a strip of road ahead flanked by high banks and skeleton trees. Behind the trees hung a heavy reservoir of darkness which seemed about to burst through at any minute and extinguish us. Here and there our lights glinted on the whitewash of a small, square house or a bigger cluster of farm buildings. Sometimes it flashed on water running down steep slopes.

In this far, Welsh countryside, light did not exist.

The trees are long gone where I stopped the car to get out and be sick that wild January night in 1947.

The huge trunks lining each side of a long, narrow S-bend were black against an entity of blackness. Branches and twigs were no more than a swish in the sky. In spite of this obscurity and my tiredness, I still recognised the place at once when I went that way again in daylight. Through the years I got to know those trees well in all their winter, spring, summer and autumn stances, for theirs was the road into the towns where were shops, cattle markets and railway stations.

But beeches develop a secret rottenness in their hearts and may suddenly crack and fall. So they had to be cut down. The same sad decision had to be taken years later for our own monsters which had stood behind the farm some 50 years, keeping the kitchen and dairy cool.

At last the driver said, 'Here we are,' and he turned in between tight hedges and through an open gate. The farm lane was not a lot more narrow or rutted than some of the roads we had already driven over. At this point in later years, I would always be looking out northwards across the field to try to see Bardsey Lighthouse flashing. This was the first 'Welcome Home' sign and none of us ever lost the thrill of seeing it on a clear night. On this night, even if I had known about it, it would not have welcomed me.

Eventually there was a widening of perspective and the hedges vanished. The car lights spread out to show white buildings squatting heavily in two parallel rows. Water gleamed on the surface between them. As we pulled up, the car lights went out and nothing was left to see but a patch of light from a small square window shining onto concrete and grass. To its right, I made out the shape of a porch.

I stood out on the yard and felt the atmosphere of this place closing around me. I felt a little tug of excitement. Here I was, as intended.

From the buildings behind me came rustlings and a stamping of feet. There was the loose-lipped sibilance of a sighing horse. And down the yard a wild, full-throated moo-ooo-oo broke out. A chain rattled violently.

A blaze of light came from the porch as a door opened and a large lady appeared, holding up a lantern.

'Well there you are, lass.'

She put the lamp down in the porch and gave me a great hug. It was like being tucked into a big, warm feather bed. I felt I was welcome. She beamed at me and took up the lamp again.

'Mind the step.'

I picked up my case and followed her into the hall.

I glimpsed stairs straight ahead, a closed door to their right. Another door was being opened to our left, which Madge went through. The light went with her and darkness pushed forward to surround me again. I followed quickly.

Inside the room, light dropped and spread in pools from two lamps which hung from hooks in the ceiling. They hissed slightly and flickered. All around, light and dark were ballet dancers, leaping and twisting, entwining and skipping apart. Always moving. Wisps of darkness swirled round the edges of the room. Lumps of it lurked menacingly behind heavy chairs. And between light and dark prowled the shadows. They loomed large then slipped from sight as people moved among them. I absorbed it all. We became part of each other, this house and I, and this room was the heart of the house.

My Uncle Cecil uncurled himself from the low armchair in the chimney corner, carefully avoiding the huge oak beam as he stood up, and came to greet me. I can't remember if he kissed me. I think we probably shook hands. He was tall and thin; very like my father, though I didn't see him as anything but a stranger at that time.

I stood waiting to be told what to do.

Something was going on between my aunt and uncle. There were glances in my direction and a muttering to which I closed my mind as it was not my business. Cecil went outside and came in again and there was more muttering.

'Have you got 15 shillings to pay the taxi, lass? We haven't got any change.'

This was, of course, long before the days of plastic cards and it was unusual for any but the utterly poor to have no money in the house. Fifteen shillings was not a huge amount of money for the average person to have about them.

'Oh, of course. Yes. I'm sorry. I...' I fumbled in my bag, mortified by my stupidity, my lack of knowledge on how to deal with such a straightforward situation as paying off a hired car.

'We'll give it back to you tomorrow.'

'No! Of course not.' Cecil took the money out. 'I didn't think.'

No one mentioned the matter again. The money was not offered back. I don't suppose they even thought of it any more – as, indeed, why should they?

The incident, though, was indicative of how they lived their lives in relation to money, a relationship I only recognised many years later.

Chapter 1

Beginnings

I WAS was not here at the beginning. Not at the beginning, that is, of Penrallt being our family farm. As far as I know, I was not here at its very first beginning either. But that could be another story.

The farmhouse was said to be over 400 years old when my aunt and uncle bought it in 1946. This knowledge was from information passed down, country fashion. None of us has ever tried to trace it back properly and the deeds and maps go only to 1874. There are clues if you know how to read them. The thickness and structure of the walls is one. The older the house, the thicker the walls. The size and trim of the structural beams is another, and the type of their joints and the fixings. The mighty oak beams that supported the upper storeys and the great chimney openings had obviously been built into the original house and had been rough-hewn with an adze. The very oldest part of Penrallt still visible, besides the walls and beams, was found in the top end of the house. This was likely to have been the original dwelling with contiguous animal housing. In its loft were untouched roof beams and trusses, all hand-hewn, not sawn, and the tie-beams were fixed with oak pegs.

Walls at Penrallt were a good three feet thick: vast structures made of field boulders dragged out for the dual purpose of clearing the land and providing building materials. Stones in the base of a wall could measure, without symmetrical shape, three or four feet on their various sides and might weigh a ton or more. One of these boulders at each end of a wall would constitute an entire width, front to back, and between these were built the inside and outside retaining walls. All of the lowest-level stones were vast, gradually diminishing in size through succeeding levels until at roof height they were quite small. As with dry stone walling in the fields, the builders must have acquired an eye for matching shapes and sizes to give the best fit and balance.

The cavity between the inner and outer sides of the wall was filled in with shovellings of slate, gravel and other rubble. Here and there slabs of slate were laid across the width of the wall to 'tie' the two sides together. This is a very simplistic description of the building process. Huge slabs of oak were put in for door and window lintels, which also served as weight-bearing supports for the upper storeys. Some newcomers who did not understand these building techniques had terrible disasters when they decided to take out the lintels in order to put in larger windows.

The outside walls of the house were made straight and fairly smooth for better weatherproofing. Later they would be rendered with cement. All other walls, including interior house walls, curved to the shapes of their component boulders producing a surface of deep and complex undulations and uneven, rounded corners. Wallpapering was a nightmare. There were no regular, matching measurements anywhere. Every piece of paper had to be cut a slightly – or considerably! – different length from every other. Even the ends had to be shaped to fit the slopes and bumps of ceiling and floor.

We were told that Penrallt, a fairly common house name in Wales and sometimes written in full as Pen yr Allt, meant Top (Pen) of the Wooded Hill. Mostly, though, people give just 'hill' as the meaning of allt. There are several words for hill in Welsh, each indicating a different shape or conformation, but our hill was wooded and had surely been so when the very first people settled on it. It is a prime site for a farm, being mostly flat, sheltered land with a good water supply. It must have been an ideal position for an early settlement too, 400 feet up, inland and hidden from the sea by the curve of the valley. This curve also hid the original village of Llangrannog, established in the early sixth century 200 feet below the farm. The steep climb from the beach would have been through thick, thorny undergrowth and this, together with the likelihood of finding only bogland at the top, would have put off all but the most determined invaders. People living in the farm settlement would have been quite safe from Vikings just pulling ashore for a quick pillage.

Little if any of the early buildings would have remained when the present group was started. Possibly some of the stones and boulders would have been there to reuse. Early roofing was likely to have been thatch or turf; then stone. The fine slate of the present-day roof would not have been available until late in the 19th century, though some of the heavier, cheaper slates like those on the animal houses may well have been used before that.

We do know of one famous person who lived in the farm before us. Well, she is quite famous in Wales, though even there she is not as well-known as she should be and certainly not as well-known generally as she should be. Her name was Sarah Jane Rees, but most people who know of her at all will know her as Cranogwen. She was a strong, forceful woman, as she would need to be to achieve all that she did in the age in which she lived: 1839 to 1916. She was a preacher, lecturer, poet and composer, winning many chairs and prizes at Eisteddfodau and travelling all over Wales and much of America preaching and lecturing. She founded The Band of Hope and also introduced the tonic solfa. The most astonishing thing about her, however, was her knowledge of sailing and navigation. She was a Master Mariner and had taught navigation to many of the local sea captains. There are some clusters of beech trees in the corners of two or three of Penrallt's fields, which we were told Cranogwen had planted as shelter for the cattle. These we always considered untouchable monuments and happily they are still there.

When I arrived at the farm, Madge and friends had already stripped and scrubbed away the top layers of the old house interior. I never saw the heavy, white satin paper that bolstered the decorum of the parlour, absorbing the damp and nurturing the creeping black fungus which ate into everything. This had been a room kept for serious entertaining, mainly of chapel ministers and the dead. Because of the damp, no wallpaper ever stayed fully attached to any of the house walls for long. It would slowly droop and sag, pulling against its diluted paste until it billowed out and then flopped over to expose the sand, horsehair and dung plaster underneath.

It must have been a very spectral experience, this slithering of white paper off the walls when there was a gathering of black-clothed mourners praying round a coffin.

The entrance hall and the living-room had been decorated in an even more bizarre way, though more practical in colour and texture. The entire ceiling, including magnificent ancient oak beams, was covered in dark brown, imitation wood-grain paper, the grain running round the timbers instead of along them. An immense amount of work had gone into removing this and scrubbing the size and old whitewash out of the wood underneath. I was not sorry to have missed these jobs.

Madge left one beam untouched over the front door, a memento to an earlier taste in decor at a time when bare wood was thought poverty-strick-en and comfortless. When my husband and I took over the farm in our time, we kept this memento, but I fear the next people

Section of tie-beam showing joints fixed with oak pegs for nails.

did not, despite being told the history. Everywhere apart from the parlour was covered in brown paint and paper, which made the basic darkness of the place, with its few small windows, even darker. It camouflaged the dirt, if you didn't look too closely, and people had enough to do without looking for extra work keeping the house over-clean.

Building and decorating materials were scarce for several years after the war. Timber was rationed and of poor quality, and paint and wallpaper were almost non-existent. Madge had a certain amount of paint and distemper from somewhere and soon we were putting washes of sunshine yellow over all the wallpapered walls. We rang the changes by dabbing splodges of colour in random patterns over the yellow. We used different colours for different rooms. There was yellow with blue splodges in the sitting room and blue with yellow splodges in the hall and up the stairs. This sophisticated high art resulted in various shades of green creeping in here and there. So then we tried mixing a quantity of yellow and blue together for a green bedroom and achieved a fine mix of shades, all drifting into each other in very artistic fashion. We still had to restick the paper quite often when it slid and drooped from the walls; which it mostly did, of course, when we had visitors. But at least we could touch up the highlights and put a wash over the black moulds. And it was amazingly lighter and brighter.

We painted all doors white, and as the walls in the hall were wooden planks, these were stripped and also painted white, making a light, welcoming entrance.

This was a time of new beginnings for a great many people. There was a widespread nostalgia for the country and a longing to get away from the drabness and hardship of war-tired towns and regimented life. Many men returning from fighting had dreams of a little farm where they could grow their own food and raise their children in peace and a healthy atmosphere.

Few of them survived more than a year or two.

They came with wrong attitudes, unprepared for the harshness of nature and for life without urban amenities. Some were people with very limited amounts of money; often thrifty working-class people, middle-aged, who had managed to buy a house and now transferred the mortgage to a farm. They chose these far-off, rugged places like West Wales, Yorkshire and Cornwall because farms were cheap. Besides being remote, they had poor quality land. This made it much harder to work and get a living from.

Better-off people had money to spare after their basic

Adze – an old tool used for shaping planks and beams and for levelling their surfaces. The knob at the back of the blade was used for hammering the wooden nails into planks.

investment but they tended to spend it on improving their houses when they should have been ploughing it into land and stock. Often they came with a little knowledge gained from books and agricultural manuals. They assumed this to be superior to local knowledge. These people mostly failed, cut their losses and left.

The poorest people who failed did not have this option. They were tied by a mortgage they could never pay off, their only income a small and unreliable milk cheque that barely covered their most fundamental costs. Life was hard and mucky and isolated. It was worse than town slums because there were no close neighbours and no nearby relatives to give help and companionship.

Penrallt was not a very remote farm and its new owners were neither poor nor unaware of the priorities of building their business and learning their trade. My uncle was a research chemist who had initially studied food and diet. He understood plant constituents and their relation to health and diseases. Despite all his academic knowledge, however, he accepted that he knew virtually nothing about farming and he was always eager to take whatever help and advice he could get from local people.

If well run, the farm could produce enough food and money for a comfortable, basic living. And it was well run, with the help of knowledgeable farm-hands, advice from the old owner and other neighbours, plus Cecil's understanding of how things worked and Madge's organising powers.

But Madge and Cecil were not really into the rigours of basic living. This whole project was almost a hobby, a way of enjoying life rather than being a serious business. Money somehow slipped away.

I thought my aunt and uncle were really rich. This was not so much for what they spent but because of their apparent complete indifference to the comings and goings of money. It was many years before I realised this attitude was not because they had so much they didn't need to think about it. They just saw no point in thinking about it. Had they done so they would have had to accept the fact that they hadn't got nearly enough and must try to live accordingly. No fun in that!

They were fortunate in being blessed with a number of elderly, rich relatives who died at strategically opportune moments allowing a certain number of debts to be paid off and a few more luxuries to be bought.

To me it was all luxury. We did work very hard but there was lots of time in between for fun. It was mostly free fun: walking, swimming, picnicking, talking, listening to music. Interesting people came to stay and I was suddenly one of them all and no longer a child. I had no trouble at all in adapting to this new life.

Chapter 2

Language and Sayings

THE local people were kind and friendly. When Madge and Cecil bought the farm they were welcomed, but there were some critical remarks made about money and culture, especially from a few strong nationalists.

'You rich English come here buying up our best farms and spoiling our language.'

They were right on both counts. Our farm was the best in the district. It was predominantly flat land with a reasonable depth of fertile soil. It was sheltered and well drained and had a plentiful supply of fresh spring water that was reputed never to dry up. Six acres of flourishing woodland provided an abundance of useful timber. There was a range of well-kept, conveniently assembled buildings, together with the house, all built on a large flat yard of solid bedrock. Adjoining the barns and stables was the haggard or stack yard, a flat area of about one acre where the harvested corn and hay and threshed straw were stacked.

There must have been many local people eager to buy the property, if only they'd had more money available than my uncle. It's quite likely that some did have as much and more, for in common with all farming areas, this very poor part of Britain had become much more prosperous during the war. But it takes time to acquire rash spending habits after long and dire poverty, and Cardiganshire was one of the poorest areas in the whole of Britain.

The other charge of corrupting language and culture was also indisputable. After centuries of English dominance and oppression, you could still walk into any pub or shop and hear this beautiful, singing language with its soft vowels and its rolling double lls and its seemingly every other word a 'bugger' or a 'bloody'.

Welsh is often called a difficult language to learn, but it seems to me no more difficult than any other of the main European languages, and certainly much less complicated than German, Latin or Classical Greek. It does have one particularly difficult foible, unknown in most European languages, of changing the initial letters of words, sometimes to indicate gender or to follow a preposition. This is called mutation, and though it is complicated, it has only to be learned. The real difficulty, faced by English people across the globe, is that the locals can speak English and most do so very readily in 'mixed company', though in 1947 older Welsh people did it with difficulty. It was their literal translations of the Welsh into English that gave rise to the stereotyped Welsh idioms used in jokes and in the speech of film characters.

Madge had a joke, an impersonation she did of the station announcer at Carmarthen railway station in the early years of our immigration. Madge reckoned the lady announcer had very bad adenoids and she would give an exaggerated imitation of her saying, 'Garbarthig, Garbarthig.' In fact the lady was probably saying perfectly clearly and correctly that the train had arrived in, or at, Gaerfyrddin (pronounced Gyrvurrthin – hard G, y as in fly, th as in these). The unmutated words would be Caer Myrddin, meaning Merlin's Castle.

Another of the stories Madge enjoyed telling was of an early telephone encounter in Wales when somebody chatted to her for some time in Welsh before asking in English, 'Are you Plwmp?' (W is a vowel in Welsh, having a short 'oo' sound, as in took. Plwmp is the name of a village a few miles from Llangrannog.) Madge claimed to have replied, 'Well yes I am rather, but I don't see it's any of your business.' Though I always suspected this response was more in the mind than in uttered words.

Most Welsh people were chapel-goers: nonconformist Methodists or Wesleyans. Chapels and churches were always full in those days. One purpose of the right of way through Penrallt was for people to walk through to the chapels. Many came in wellingtons and changed into shoes when they reached the road, leaving a pile of boots in a corner of the lane. The perpetual wearing of wellies, often with bare legs for the women, for trousers were not worn, produced a dark ring round the calves. This was partly ingrained dirt, partly the wearing-off of hairs. It was amusingly incongruous with smart black chapel clothes, shoes and stockings. We all had the marks, men as well as women. They only disappeared when everyone had baths with plenty of hot running water and plenty of time to soak and scrub.

We were also amused to see that husbands and wives who kept shops or pubs tended to have different religious inclinations, one going to the Wesleyan Chapel, the other to the Methodist. This allowed members from all congregations to happily spend their money where they liked without disloyalty to their respective faiths. Care was taken not to give offence. Mr Thomas, Siop Canol, enjoyed an occasional discreet bet on the horses and the year Sheila's Cottage won the Derby he came quietly into the shop from a back room and mentioned casually to Mrs Thomas that 'the young lady from Penrallt has just come in.'

Nobody ostracised us for our lack of religious observance or our social permissiveness. They were cheery and matey with Madge and me when we went into pubs, even unaccompanied, though none of their women were ever there. In the early days, women who came to our house on a private visit or for a Women's Institute meeting always refused the smallest sherry with a modest, 'Oh no, I don't drink.' It was therefore a revelation on one WI outing to distant parts, to see many of these ladies knocking back straight whiskeys with every sign of enjoyment and competence.

A certain number of men, because of positions of eminence in chapel, school, local government and the like, did not drink either. These were admitted discreetly to small back rooms in public houses where nobody saw them go in and everyone knew they were not drinking anyway.

When Madge and Cecil first moved in, one of the village dignitaries gave Madge a welcoming lecture along the lines of, 'You'll find us very friendly, always ready to help and give advice. We may call you heathens and pub-crawlers but you are welcome and we will pray for you.'

It was the pub-crawling bit that stuck in Madge's gullet. As she said, 'There are only two pubs in the village and you can't favour one over the other. We do call at Brynhoffnant sometimes, on our way back from Cardigan. Not much of a crawl really.'

One of the very nicest sayings, which Madge enjoyed very much and delighted in repeating, was the greeting: 'Welcome home!' It was said even if you had been away for only a day and always with a warm sincerity that gave us a pleasant glow of belonging – even though that belonging was a little beyond the pale.

The saying 'mochyn dau-dy' – 'a pig of two houses', was used of people who cohabited with more than one partner. Mostly these situations were treated with mild disapproval or amusement, but there was one shocking affair that provoked great concern, anger and retribution. A married minister from the chapel in a neighbouring village took to creeping into his lady friend's house late at night. He could not have supposed he was unnoticed. Even I, after only a few months in the district, knew that nobody could ever do anything at any time around here without someone seeing them, or somehow knowing about it. He was warned off, but still persisted. Eventually the most morally outraged men, and possibly some who were just mischievous, gathered outside the lady's house in the middle of the night when the minister was known to be inside, and nailed planks over the doors and windows. The minister was not let out until well into the day, when he had to walk away in shame in front of everyone. The poor lady involved virtually never went out in public again.

Another pig saying was a drinking toast: 'i bobl fochyn tew', meaning 'a fat pig to each of us'; or literally, 'to each a pig fat'. (You will have recognised a mutation here from mochyn in the first saying to fochyn in this one.)

There were many family sayings that went on through time even into my present-day family. Of Madge it was said, 'Give Madge two people and she'll organise them.' But Madge also possessed the rather rarer art of being able to achieve this in such a way that most people didn't realise they were being organised at all. She did a very good job in this way with the village committees, as almost everybody who lived there at the time agreed.

A family saying still in use in my family is, 'While you've got your boots on...' It can be used in the context of being asked to fetch something from upstairs after you have been up and are now down again, or any situation like that. It arose from Madge calling out to whoever had just come into the house, usually moments after the troublesome business of removing their boots had been completed, to just do some chore outside like 'shut the chickens', or 'bring some logs/paraffin/extra milk'.

Several sayings referred to eating: I'm FUFTB – Full up Fit to Bust. And there was a rhyme:

> The Lord be praised
> My belly's raised
> An inch above the table
> And I'll be damned
> If I'm not crammed
> As full as I am able.

Cecil and I were both poor conversationalists, so when we went out in the car together there were long silences. This didn't matter when we got to know each other, but in the early days he felt obliged to say things from time to time. And so he read out road signs as we passed them. Only he didn't just read them, he turned them into backslang. It took me a little while to work this out, for he spoke quietly, sometimes indistinctly. When I did though, I was delighted, for my father and I had talked together like this. It was not just a simple exchange of letters but whole sentences were shuffled about. A very simple example that comes to mind – for I have largely lost the skill now – is: 'Oo willy sarp uhm pay eece cof ekal eapse', which is 'Will you pass me a piece of cake please?'. I taught my children and they keep it up, though mainly with single words. It needs constant practice to be agile and I am astonished to remember how I spoke these words and sentences straight from my head almost like basic language.

Cecil used occasional German words and phrases in his conversations, gathered from pre-war holidays. Germans have two words for eating: one applies to humans only (*essen*), the other must be used for animals only (*fressen*). Cecil would insist on using this latter word in all circumstances, partly because he thought the distinction ridiculous, and partly to annoy Emmy. Emmy was a German refugee from the mid-1930s who went to live with Madge and Cecil and became another (the first I think) member of their extended family. When she first went to Llangrannog beach she stood enthralled by the great rock, Carreg Bica, that stands between the first and second bays.

'I like that rock. I do like that rock.'

'Well, you can have it!' said Madge. And it was known in the family as Emmy's Rock ever after.

As well as losing its name, this rock also, more publicly, lost part of its substance when in October 1982 a section slid out of its middle. Somehow the whole thing settled down again and has remained in mostly one piece ever since.

I don't know when I first acquired the name Sheila Troednoeth (pronounced, roughly, troydnoyth). It was certainly within the first year or 18 months, for I was going barefoot very early on, and that is what it means: Sheila Barefoot. There are still a few people about today who know me by that name.

Cecil had a great many sayings, mostly from rude jokes.

'You're rushing around like a fart in a colander.' And when there was a query about what to do with something, 'The first man who tells her goes straight back to the ship.' I laughed at this because it seemed a funny thing to say. I was told the story of the old lady who called at the Mission for Seamen, saying she would love a few homesick sailors to come to her for a proper Christmas. Four or five were 'volunteered' and the Petty Officer had to go with them to ensure they behaved themselves. They all had a lovely time putting up holly and mistletoe and decorating the tree. 'Now,' said the old lady, 'what shall we do with this last candle?' And in jumps the Petty Officer with the punchline. I laughed hilariously – but I understood the meaning of the joke no more then than I had before I heard the story!

Madge was talking about someone in the village one day, trying to describe to Cecil who she was. 'She drives round in a little (whatever car it was). Lives on the back road.'

'Oh! That's the woman who knows where her wheels are.'

And Little Mrs Jones Who Knows Where her Wheels Are she became ever after. She was very pleased when told. Thought it a great compliment, as indeed it was.

In 1934 or 1935, when he was a research chemist at Woolwich Arsenal, Cecil was asked if he would go down to Aberporth, which is on the coast some five miles south of Llangrannog, to be in charge of the rocket research station just being set up there.

'What!' he said. 'I don't want to be stuck in a dead-and-alive hole like that!' Yet now here he was, in very different circumstances to be sure, not the least of them his dress style. He wore shorts for a lot of the year, mostly several sizes too large because he hated constriction round his stomach. In hot weather he rolled the tops of these right down and the legs right up, and being very thin and very brown he looked rather like Gandhi. Susan, when small, said to her mother one day, 'I don't know how Cecil's shorts stay up. They're hardly on his bottom at all.'

When Madge and Cecil moved in, an obvious enquiry to be made of the locals was, what's the weather like? 1946 had been one of those 'wettest summer since Doomsday' years, so was obviously not typical.

Well, a bit wet it is, they had said. When you see the mountains clear across the bay, you know there is rain coming. When you can't see them, it's usually come.

'What mountains?' said Madge.

'It was Christmas before I saw them,' she told me later, recounting this anecdote. 'Till then you couldn't even see across the farm most days.'

I was talking recently to an old friend who had been a little lad working at Cefn Cwrt Farm at that time, Geoffrey Probert. He told me he'd just been to Australia where he had gone on a coach trip to see Ayers Rock. This is a place where it almost never rains. At least, nobody ever sees it rain. And Geoff said they would see it rain that day. The coach driver said there looked like being a thunderstorm but there'd be no rain. Geoff tried to bet them 50 dollars but no one would take him. And it rained!

Carreg Bica. This beautifully posed photograph of Llangrannog's ancient rock creature was taken by Beti Sendel, Penlon, Llangrannog. My thanks to her for the picture and for permission to reproduce it. The rock is known in Collins myth as Emmy's Rock.

'How did you know?' they asked, amazed.

'I come from Wales,' he said.

'I could smell it,' he told me.

Another key question Madge and Cecil asked was, 'Do you have much snow here?' The answer provided another of our long-standing family jokes:

'No. No snow to speak of.'

And indeed, after one of the wettest hay and corn harvests of all time, Christmas and New Year did saunter in with warm sunshine and the promise of an early, bright spring.

In this sunshine I explored hills and cliff tops, I paddled, and I climbed over rocks and round cliff bottoms. I had always loved the sea and high places, and here I had both together. Heaven.

Chapter 3

The Amenities

I REMEMBER nothing of the rest of my first night at Penrallt. I suppose I ate and drank something. I slept in the little room and bed which were to be mine for most of the years I spent with Madge and Cecil. And I must have had the toilet arrangements explained to me. (I think toilet was a new word I acquired at this time in place of my accustomed word, lavatory.)

Toilet arrangements were primitive but complicated. Even at this early stage I was told that only in the most awful weather or direst emergency did one use the inside arrangements. I have no doubt I went out before bed that night, in my high heels, to find the grass patch up the lane, as though I did this sort of thing all the time.

I found out almost immediately that my new relations were nudists. Far from regarding the body as a rude and nasty thing to be hidden away and never referred to in thought or word, they flashed their own around as though they were normal, everyday parts of life. There was no way I was going to be able to live here without doing the same.

Away went 17 years of tense, giggling embarrassment and inhibition, almost without trace. I behaved as though this was the only sort of life I had ever known. I nonchalantly stripped off to wash or swim among people I barely knew. Beaches that are now packed with people for most of the year were then empty. Only Llangrannog beach was thronging with visitors for seven or eight months of the year and local people wandered around at all times. We did not, of course, intrude here in our nakedness. But there were secluded coves to walk to and two miles down the coast, at Penbryn, was a vast stretch of sand where we rarely saw another person, even in the height of summer. A few naked bodies could easily blend into the background. These isolated conditions lasted only two or three years at most. Then cars and petrol became readily available and people flocked to all the beautiful, remote places which immediately ceased to be remote but were still beautiful; the more so, in a sense, since more people enjoyed them.

Very few of the people who came to Penrallt indulged in this sort of carry-on. Most of them surely had no idea of its existence, for again, it was not flaunted. Even Cecil behaved with reasonable decorum – apart from the way he wore his shorts in warm weather, when the top was rolled down and the legs rolled up to produce a very insecure loin cloth. I realise now that I never actually saw the many naked bodies I mixed with. I never looked below the faces! Only when I assiduously presented nudity as normal to my children and they rejected it out of hand, did I begin to suspect that I too had never really accepted it. Going without clothes and treating bodies and their functions as suitable for communal unconcern had remained a much more alien concept than I realised. It was many long years before I was able to understand the logic of Cecil's story about the young woman, compromisingly revealed with nothing on, who simply put her hands over her face. Even after years of exposure to naturism, I think I never evolved mentally to the stage of being able to react in that way. My hands would still have gone anywhere but my face.

Washing facilities were a wash-basin on a slate slab at the end of the long kitchen. There was none of the old 'down as far as possible, up as far as possible' prissiness here. People stripped and washed the lot, often while the hustle and bustle of breakfast preparation went on around them. Beyond this 'bathroom' was the coldest room in the house, where milk, butter, eggs and meat were kept. People already washed and working would go rushing through from time to time to fetch whichever of these necessities was required. The kitchen's basic structure was freezing, but the movement of these bodies going past, plus the opening and shutting of the dairy door, caused a wind-chill that would, according to my uncle, bring terror and mutilation to any number of brass monkeys. A saying I laughed at without the remotest understanding of its meaning.

This kitchen originally was about 14 feet long and seven wide and it contained one window, 18 inches square, in its end, outside wall. Early on Cecil put a door, half glass, in place of this window, giving easy access to the kitchen garden, and he moved the wooden wall at the opposite end of the room three or four feet along into the dairy. This provided extra space and another, biggish, window. Light from this was restricted by the three-foot depth of wall into which it was set, yet it still made a considerable improvement to the atmosphere and visibility.

The kitchen and dairy walls had been whitewashed every year but the place had been too dark to show up the accumulations of dirt in remote corners. Madge spoke of layers of thick grease behind the kitchen door, where, before I arrived, there had stood a long, wooden bench with an enamel bowl on it. This had served as kitchen sink and wash basin. As there was no back door, the slops had to be carried right through the house and cast away outside somewhere.

Madge was prepared to put up with many privations and deprivations in the name of Utopia, but this state of affairs was definitely not one of them. A sink was rapidly installed, complete with draining-board and a drain. It was a luxurious convenience after making do for a few weeks with the bowl and bench. The drainage system did away with one lot of carrying, but fresh water was still carried into the house for many years to come. Water was never scarce, but time and energy for carrying it were.

Cooking arrangements, and therefore meals, were primitive and required a lot of understanding and planning. The traditional kitchen range needed traditional knowledge and traditional recipes. The fire could be roared up to produce a hot roasting oven, but you had to learn to control it and judge its temperature. Stews and casseroles, slow-cooking rice and other puddings could be easily organised, but Madge was accustomed to more variety in her food and it wasn't long before she achieved this, despite food rationing and lack of the more exotic foods which, even in wartime, were to be found if you knew how and lived in the right place.

Another important function of the fire and range besides cooking, and keeping us warm, was to provide hot water. A huge kettle swung above the hot coals, permanently boiling or just off the boil, which was used mainly for tea and cooking. It hung on a strong S-shaped hook and could be made hotter or cooler by hanging the hook higher or lower on the long heavy chain suspended from somewhere in the vast chimney. The side plates of the firebox were the hot walls of the oven on one side and the water boiler on the other. Water was constantly being used from the boiler and it was most important to keep it topped up, both for use and because the boiler would crack if it went dry. This water was used for washing: bodies, dishes, cows, milking buckets. Body-washing water might double for floor washing.

A small cauldron of stew, bacon or vegetables could replace the kettle for a while, hanging on the hook and chain over the fire, and it or other saucepans might sit on the fire itself. It was possible to cook a decent cake in the oven, but mostly they turned out heavy. This did not suit Madge's sophisticated cuisine at all, but there was a more readily adjustable, paraffin-powered oven. This had wicks which you could turn up and down and a thermometer to tell you when you had the temperature right. Paraffin was scarce, probably rationed, but it was available in bulk deliveries to farms, being used for lighting, heat, and the running of machinery and tractors. But it was temperamental. At least, the household devices which consumed it were. They all of them, except the vapour lamps, worked on the principle of a wick soaking in paraffin which you turned up or down to control the heat or light. But they had to be watched and checked regularly.

The paraffin oven was a mean, malevolent thing. If you were in the room, you might see tiny black flecks starting to float out and about the room, at which stage you would rush to turn the wick down. If you were not in the room, the proliferation of soots would be exponential. It was as if the creature put out these first small specks to test its chances and then belched out great volumes to do as much damage as possible before someone came back and caught it. I remember once being unable to see into the room at all, the air was so thick with soots. It was horrible to clean: greasy and sticky. It was food and soul-destroying. I learnt many strange new words on occasions like these.

Poor Madge did her best to understand and control this cooker, but it was always when the most special cakes or dinners were cooking that some emergency occurred outside and all hands would have to rush to help. If a moment were snatched to turn the wicks down, the monster could still win by going out. After six months of burnt or soggy cakes, a civilised gas cooker was brought in, complete with a large cylinder of Calor gas plus a spare to avoid the disaster of running out. Madge's meals could now return to something approaching her old gourmet standard.

Drinking water was pumped and carried from about 10 yards down the yard using five-gallon tinplate containers. These were shaped rather like miniature milk churns. They were not heavy but must have been very strong, for I never knew one to leak or lose its handle in all the years that we used them, and they were old before then.

Three of them stood in the kitchen and whoever emptied the second one was supposed to then take the two empties and refill them, it being altogether more balanced to do two at a time. In practice, this was something Madge almost never did. She was not really built for water carrying, being short and bulky.

Washing water came from a well across the yard, opposite the front door. The well was about 1½ft deep and maybe 3ft x 3ft square. Above ground it was surrounded on three sides by walls about 2ft high with a solid slate slab laid on top of them. This kept the worst of the wandering rubbish from getting in but made getting water out very difficult. You had to lean forward under the slab and stretch down to dip the can into the water, afterwards lifting the heavy load up in an awkward, twisting movement. This heavy unbalanced lifting eventually did something nasty to my back when I got stuck with the can half out of the well. I was in this position for at least five minutes, though it seemed much more before I could move. After 20 years of constant pain in back and legs, I was able to visit an osteopath who eventually cured the problem.

Water from this well was carried in buckets for the cows to drink when they were tied in their stalls in the winter. There was a standpipe at the back of the cowshed, but it was quicker to carry from the well, especially as tap water came from the tank and had to be pumped. Pipes and automatic drinking bowls were fitted in the cowshed four or five years before we had water on tap in the house. Not that anyone begrudged them. Carrying water to the cows was much more troublesome than carrying it to the house.

And so, of course, we had no water closet.

Our Ty Bach was known as the White House, partly, I suspect, because of Cecil's extreme anti-American sentiments, but more obviously because it was painted white. It was right at the back of the house. To get to it we had to go out the front door, round the end of the house and through two gates, a fair trek, especially in cold wet weather. After the door was fitted from the kitchen to the garden, this trek was halved.

It was a chemical toilet, Madge and Cecil having decided that this was more acceptable than untreated sewage. It all went onto the farm muck heap and eventually out on to the fields. I don't know how it affected the rest of the manure. It was said to be safe but it was quite fierce, the chemical, and had to be treated with care.

I was in the kitchen with Madge one day, quite early on in my visit, when a crashing and bellowing as of a rogue elephant, one could imagine, came towards us from the front door.

My uncle pounded in.

'Cecil! You haven't taken your boots off!' Wellies were not allowed in the house.

'Bugger my boots. Give me some water quick or I'll have my balls off.'

And he let go of his clutched waistband and allowed his shorts to fall round his ankles. He had no underpants.

'Cecil!' Madge was used to Cecil's strange ways but never failed to show outrage – in public, at least. 'What do you think you're doing?'

'Bloody Elsan's splashed me. It's burning my balls – '

'Yes, all right. You needn't have come in the kitchen.'

She sent me off to collect the eggs, or some such task.

It was probably only a small splash but he was a great exhibitionist. Madge would tell the tale of how she had to struggle to prevent him from removing his shorts outside a busy railway station one day because a wasp had crawled up his leg. She didn't succeed but did find a coat to conceal him behind. He made a great song and dance about that too. A normally very reticent man, he was determinedly pioneering about freedom for bodies.

Peeing in the Elsan was forbidden. In all but the most inclement circumstances we joined the beasts of the field for this necessity. This kept the Elsan from filling up too quickly, which would have wasted chemical and time. Surprisingly, almost everyone who came to stay – and many of them were from quite high layers of society – accepted these toilet arrangements with barely a hair turned. Any field corner or secluded nook might reveal a squatting bottom or a man gazing contemplatively at the view.

For comfort in the winter and on stormy summer nights, a second Elsan was installed in a dark, tunnel-like, junk room above the lean-to dairy. The floor was of wooden boards, which also served as the ceiling of the room below. Long after any but the very oldest of us recalled its origin, this room was still known as 'The Elsnanole'.

Beside this toilet stood the Pee Pail, a galvanised bucket, large and vibrant. Piddling in the Pee Pail produced a timpanic trill that would have enhanced the sonics of many a modern musical composition.

The sound came clattering one night into a social gathering. Even to those not intimately acquainted with our unusual conveniences, the tone and rhythms were unmistakable. Doors had been left open and the fanfare was performed with deliberation and panache. Madge called out, partly in bravado, partly in the faint hope that some acceptable explanation might be offered:

'Cecil. What are you doing?'

'I am peeing from a great height.'

Chapter 4

More Family Members

THAT first morning I didn't know where to start among all the exciting prospects on offer. I went outside and stared enthralled at the 'mountain' across the valley.

'I'm going up that!' I said.

But before I did anything else at all, Cecil was taking me to meet the animals, the Funnies. On the way, two dogs allowed us to talk to them. These were Jacko and Belle. A third dog, Fan, watched and cringed from a distance. Fan was a very beautiful, pedigree border collie. She had long silky hair that was a deep auburn colour with some white patches. Her eyes were also a light gingery brown. She was a very nervous animal, always difficult to approach. Belle was a wall-eyed dog, that is with one brown eye and one blue. She was a fairly ordinary collie, black and white and friendly. These two working dogs came as part of the farm stock, though I think they did little work for us. We were not really into working with dogs. Jacko had probably been a good cattle dog too, as well as a rat catcher. He was Griffy Jones's special dog, left behind because the old farmer thought he would pine in a village house with nothing to do. He was a strange, cartoony-looking dog; something like a heavy Jack Russell but with huge,

An armful of kittens.

turned out feet from being crossed with a corgi. Numerous cats and kittens peeped at us and ran in and out and around the buildings.

Two parallel lines of buildings stretched down the yard towards my mountain. It was like an ancient street with huge old stone houses each side. The road between was rough, ridged stone and there were hollows full of water. This road sloped down towards a gate beside which was a high heap of steaming straw and muck. A trickle of brown liquid ran out of the house opposite the farmhouse and made its way by numerous cracks and channels down the yard and into the muck heap.

Cecil and I went first across the yard and into this house, which was the stable. Chunky cobblestones were laid at the entrance as well as on the floor inside. Any other surface would have broken up under the heavy stamping of iron-shod hooves. The room was dark and steamy. The smell of warm urine and dung wafted over us. I didn't think it unpleasant. Two huge, chestnut-coloured bottoms loomed up in front of us. Cecil walked forward and patted one.

'Hello, Bess darling. Here's your Aunt come to see you.'

In theory and in stories, I was mad about animals, especially horses. But Black Beauty, and the Cowboy's Best Friend, even the milkman's horse in town, tied into her shafts, were quite different from this reality. I stood well to one side of the great legs and feet and stretched out a tentative, long-distance pat.

I leapt back as she moved, but she was only swishing her tail.

'Don't tickle her! She thinks you're a fly.' He gave her another couple of gentle thumps, then did the same to the horse in the next standing. 'This is Star.'

I did better with patting Star.

At the end of the stable were two young speckly-grey horses called Punch and Judy. They were unbroken, and after running loose during the summer were sold, along with Star who was an old horse. And yes, there was a vague concern among us that something far worse than the usual, reasonably humane, death would happen to these animals, but the morality of farming was not much thought about at that time. Such thoughts have worried me since, as they have also grown among the general public, but it is too vast and complicated a subject to feature much in this book, leading on as it must to so many other moral problems. Bess stayed with us another five or six years, doing all sorts of useful work about the place. She was then quietly shot on the yard and went on to become dog food. And what better way for any of us to go?

Rural West Wales was one of the poorest areas of Great Britain, largely because it was so far from the big towns and ports. Roads were bad, transport difficult and expensive.

Modern technology was less in evidence here than in more accessible parts of Britain, but even here, tractors had started to replace horses. Farmers were being urged to produce more and more food and tractors worked faster than horses. In any case, the skills of working with horses are built up over many years and Cecil had neither the time to learn nor the strength to perform them. Engines he understood and could control, so he bought a tractor and a milking machine.

I was taken to the cow-house next, to meet the Ladies.

There were only 12 milking cows on the farm, plus half a dozen calves and heifers. This was the average for this type and size of farm at that time. It allowed about six acres per cow. When you consider that in 1980 we were running four to five times this number of animals, you will get an idea of how production was pushed up. Modern levels of stocking are more like one acre per cow.

Writing this now, I am very pleased to find I remember the names of six of those original 12 cows.

They were Cardi, Cochen, Seren, Rose, Brocken and Black-and-Little-White. Or was that White-and-Little-Black? Anyway, she was quickly renamed Whiskey. They were all with us for several years, which is probably why I remember their names and forget the others. The Bill of Sale mentions no Brocken, but has Gwen, Dolly, Molly, Rose 2nd and Miss Phillips. I remembered the names Molly and Miss Phillips when I read them. I didn't do much milking for the first year or more, and during milking is when you get properly acquainted with cows. Perhaps they weren't with us for very long.

I remember Rose going quite soon after I arrived, because I wrote a poem about it. In the Bill of Sale it says that she was a third calver and carrying her fourth calf, which would make her about six years old. This is quite young to be culled even a year later, so there must have been something wrong with her. I remember Cochen too, because she was a great character and must have been one of the last of the oldies to go in 1953. She was a second calver and carrying her third calf in 1946, which would make her between 11 and 12 years old when she was sold. Not a great age for a cow in those days. My husband and I kept one cow till she was 20. She was shot on the yard, as were one or two other elderly ladies who had served us well. It was not a kindness very poor farmers could afford.

It was always sad when animals had to be sent away. We were fond of them and sorry for them, but farming was an established way of life and if people are going to eat meat, drink milk and wear leather, then animals have to be killed. There was less pain and stress for most animals in those days as there were a great many small slaughterhouses

Cardi coming in to meet the test-tube bull.

only a few miles ride away from any farm, and so the job was quickly done. Every bit of an animal was used for something then. Even the toughest of meat was eaten. People were far less fussy, wasteful and greedy than they are today – though only, of course, because they had to be. A return to those old ways would be great for world conservation.

In the cow-house we were met by a row of heads. They were great big heads and they were hanging over a wall, shaking and nodding, chewing and dribbling. Cows all had horns then. Very few have today. And they were so varied: long ones, short ones, some

The enlarged herd in later years, with no horns. Still lovely assorted colours.

broken, some twisted. They grew upwards or downwards or sideways. Sometimes one was up and one was down on the same cow. Cows could do great damage with their horns, mostly to each other.

It was the noses I noticed first though, not the horns. Huge, shiny, slimy noses. Cartoon noses with cavernous nostrils. Every now and then a tongue would thrust out from a hidden mouth, curve upwards and delve blissfully into one of the caverns. After a moment it flicked out and slid round to clean the second nostril, slipping away into its mouth again then, like a snake into its hole.

Cecil rubbed his face against a cow's face. Its great tongue came out and licked him.

'Have you got a kiss for your new aunt, Rose darling?'

Obediently I moved within range.

'They're a bit rough. Like a cat's tongue.'

I put my cheek close to Rose's and her tongue curved out towards me. There came a rasp like heavy-duty sandpaper across my face. I thought all the skin had gone and I jumped back with a yelp before she could scalp me.

Cows don't bite grass off in the fields. They have only one row of teeth and these are in the bottom jaw. They use their rough tongues to wrap round blades of grass, grip them and tear them off. The tongue then works with the teeth and the hard roof of the mouth to break down the cellulose in the grass before passing it down to the first stomach for further processing. Then it's regurgitated back to the mouth for more mastication. It's a lot of hard work, breaking down and digesting grass, then turning it into milk; which is why cows spend so much time lying down ruminating.

There was not a bull. We had our bull seeds delivered in a test tube.

On my way down the yard, I also met David and Tom, who did the bulk of the heavy, everyday work of the farm. David was a short, solid Welshman, dark hair and eyes, very strong. He cycled each day about five steep up-and-down miles to be at Penrallt for 7am milking. I don't know how old he was but he was married with at least two children. It was David who kept the daily and yearly routines going in those early days, who taught us the ways of harnessing and working the horses, of putting up a good load of corn or hay, of knowing the times and seasons for all the jobs of the year. Tom, who helped around generally, was a boy of about 16. He was also Welsh, but tall and fair, I think, and quite slender. He lived on a farm a mile or so up the hill. Being one of several children, he could be spared to work for money away from home. He was quiet and perceptive. Perhaps he had a wider education than David for he was more aware of other ways of behaving. David was intelligent, clever, hardworking, polite. Altogether an average, nice person. But his table and eating habits were of a basic, no time to hang about type. He never did seem aware of the horror that hung in the air as he blew down the silly little hole in the salt cellar when the salt refused to come out. For that matter, neither did Madge ever condescend to put it in a more accessible container where the all-pervading damp would be less troublesome.

At the end of 1947, just a few days before Christmas, David had an accident as he set out for work and he was unable to get to the farm. I was away at this time. It was the first of my two flights back to civilisation when I felt I should be doing something more constructive in life. The second was in 1952. I have a letter from Felicité, though, which tells me all about that Christmas on the farm. Felicité was the second in what might be called Madge and Cecil's adopted or extended family. Unlike Emmy she did have a perfectly good family of her own, but she moved in with Madge and Cecil during the war when, like them, she was dispatched to Bridgend on war work. Later she spent a lot of time at Penrallt on holiday or in longer breaks between jobs, and she got a taste for the life. She took a job on a friend's farm in Devon for about a year in 1951, eventually moving back in 1952 to become a permanent part of Penrallt and later its part-owner.

Anyway, in 1947 Felicité was having a prolonged Christmas break while she sorted out her next career move. In her letter she tells me of all the catastrophes that occurred, starting with David's accident. Cecil was also ill over the same period. He always had a lot of health problems. Flicit and Madge had to deal with the milking and all the essential farm jobs, as well as keeping the house running. Cecil was up and about again in a day or two, but it was still a lot of heavy work for him and Flicit. All the Christmas visitors mucked in when they came, doing what they could to help, and no doubt Christmas dinner was a great success, as always. The visitors would have brought wine and spirits to swell the very limited rations available. Flicit said she didn't know how she would have survived without the odd snifter. The Joneses at the Pentre Arms had promised a bottle of gin, and luckily that had been picked up before the emergency because she says, 'believe it or not, we didn't manage a single visit down there over the holiday.' Cecil became very ill again after Christmas with gastric flu, but they managed to find a man in the village who could come and help out.

It was soon after this that Flicit got her new job at Cambridge Low Temperature Research Station and, wondering what on earth they'd do on the farm without her, I decided that I really did have a perfectly good and proper job to do there myself. I knew I would be very welcome and useful and happy, and so one lunchtime in the second week in January, I suddenly turned up. Three faces stared at me in some astonishment. I can't remember what Madge said. Cecil said, 'Good-Gawd-all-bloody-mighty!' and someone said, 'This is William Henry'. Madge probably quite soon said, 'Welcome home' and 'Get yourself a plate.' I said 'Hello William Henry', no doubt got myself a plate and sat down. I looked around the table then asked, 'Where's David?' I was told David wasn't there any more, and again, 'This is William Henry.'

And so I met this William Henry who first of all helped out when David had his accident and then became a permanent worker in place of him. Poor David never properly recovered from his accident. He had run into an old man who subsequently died, and although it was established that neither the accident nor the death was David's fault, it must have been a very nasty trauma for him. William Henry, mostly known as Henry, fitted in very well to life at Penrallt. He was a very nice, gentle man, but with a bit of a twinkle and a quiet sense of humour. I was very fond of Henry, and of his wife Morwen. She still lives in Llangrannog, one of very few remaining of the old originals. Henry sadly died many years ago. Another of our family sayings came from him: 'Oh well, any port in a storm.' True, it was already a saying, but Henry gave it an extra twist, commenting on the ducks trying to mate in a puddle on the yard. The drake could never

David cleaning out the surface well.

get the job done because he kept falling off the duck's back. He couldn't balance without a proper depth of water.

The rest of the great stone buildings down the yard that weren't housing animals were full of huge machines and bins and sacks and buckets. They had dark, rustling corners. Sometimes you would catch a glimpse of a rat scuttling across the floor or along a beam, or sitting up washing his whiskers. Behind the barns were stacks of corn to admire and hay to climb on. Ducks, chickens and geese ran and scratched and sat about everywhere.

Right down the bottom of the main yard, round the corner opposite the muck heap, were the pigsties. Pigs were a particular favourite in our family and I was pleased to scratch their broad backs and exchange a few friendly grunts with them. Huge they were, and round. Rather like barrels with ears. Like monster money-box pigs, in fact.

'That's Claud,' said Cecil of the nearer one. 'He'll be helping out the bacon ration next month.'

'Oh!' I said in dismay, watching them snuffling and snorting in their straw. 'Poor piggies.'

'They've had a good life. Better off than a lot of people.'

I thought of all the awful things that had happened to so many thousands of people in the last 10 years or so, and had to agree that the lives and deaths of these pigs would be incalculably happier and gentler, however it was done.

I looked over at the pink and white barrel filling the other sty and asked the question that anyone of my age reading this will almost certainly be asking themselves, remembering the radio show, *ITMA*.

'Is that one Cecil?'

'No. She's Cecily.'

I giggled, and I thought they would have more reason than the *ITMA* couple to be forever saying, 'After you, Claud.' 'No. After you, Cecil.'

I climbed up my mountain, like the ignorant townie I was, by taking a straight line across the fields. It would have been quicker by the roads, what with hedges and bogs and the stream. The stream is marked on ordnance survey maps as The Hawen. But what do they know? It is actually the Collins Piddle, marking the lower boundary of our land. Luckily I didn't meet the owner of the mountain as I prowled across his land. He was the cartoon-type irascible farmer (English), who charged at anyone he saw, even on the footpaths, waving a stick or a gun and shouting, 'Get off my land!'

It was a lovely sunny day. I enjoyed the climb and was thrilled by the height. It was exhilarating. And then I went over the top and saw the sea. I don't remember if it was flat or rough, blue or grey, or if the North Wales mountains were visible. I just breathed it all in with an immense joy I had never felt before and I knew I was going to just love it here.

Chapter 5

There Goes the Hooter!

ONE of the big events of the farming year was threshing. The stacks of corn in the haggard were opened up and the sheaves fed through the threshing drum to separate the grain from the straw. The grain was then stored in the granary to be used for cattle and horse feed. If there was enough top-quality grain, that was kept for some of next year's seed. The straw was restacked and used for food and bedding. Threshing was done two or three times during the winter months.

The threshing round started as soon as the bulk of the harvesting was over. In a good year this would be the end of September, going on through October. During this time, the threshing drum and work gang would be on the job somewhere, weather permitting. Farmers were waiting to prepare their winter food and bedding and if it had been a wet year, as 1946 was, the sooner the wettest sheaves were threshed and the grain spread and turned to dry it, the more might be saved from going mouldy. It had to be turned and shaken regularly to separate sticky grains and get air through them.

The second threshing came round in early February. David had already been to three farms on our 'circuit' and the machine was booked to come to us the following Tuesday.

Threshing was one of the big communal jobs where neighbours got together to help each other. They were exciting occasions, bustling with activity and social intercourse. They were times for people to exchange news, stories, information and gossip, a chance for younger men to get away from the restrictions of homes which were often feudal and tyrannical. There would be great games and jokes among the men, especially if there were any unmarried girls about.

At least a dozen men were needed to keep the work going, and rather than hiring a gang, as was done in parts of England, Welsh farmers exchanged labour among themselves, returning man for man within set groups. It was rare for a farmer to go threshing himself. He sent his farm servants or his sons – basically the same thing but paid even less. Each farm was the centre of a circle within which labour was exchanged. Other farms farther away were also centres, and all of these circles overlapped slightly so that each farm had a slightly different catchment area and a slightly different group of workers. It was a very simple and effective system, although remarkably difficult to describe. Numbers of men turning up were variable. They might be low, due to illness or accident at home. Or extra people might come to help, especially if there was hardship on that particular farm. Mostly, though, if you were a good neighbour you got a good crowd, especially if you had a reputation for providing good dinners. Newcomers, too, would have curiosity attraction. David thought there'd be at least 20 coming to us.

I was looking forward to the excitement of the occasion, eager to join in the work and show off my strength – an unfortunate and annoying ambition, common to many townies visiting farms, as I found out during my later farming life. Loading carts or feeding the drum, young men were indifferent to the need for skill and cooperation in their job and yours. They only wanted to prove – to themselves as much as to others – how strong and fast they were. The only way to deal with such people was to push their bales and sheaves back off the cart, otherwise it became impossible to build your load properly. There were many things to learn.

It was going to be a very busy day in the house too, judging by the quantities of food Madge was talking about preparing.

'I wonder if there'll be enough men to make two sittings for lunch. If there are, they'll be able to keep working and keep the machine running. That makes serving food easier too.'

'Madge!'

Cecil was at the living room door. It was an indication of the dire emergency of the occasion that he had dared to come through the hall with his wellies on.

'David says the thresher's coming tomorrow.'

'Tomorrow!' Madge was too shocked to do more than glare briefly at his feet. 'You said Tuesday!'

'Yes, well somebody's umpteenth cousin's aunt-in-law twice removed has died and Johnnie's had to shuffle the rota around to fit the funeral in.'

You had to be very careful what you said and who you said it to among such distant and obscure relationships. David said he didn't go to many funerals, only those of close relatives. The last one he had attended was his wife's cousin's mother-in-law. But after all, a funeral meant a few hours off work and was an occasion to dress up and enjoy company, food and singing. The more relations you had, the more occasions to celebrate, whether weddings, births or deaths.

'If they don't come tomorrow,' Cecil said, 'it'll be another two weeks before they get here. And you know we're nearly out of corn for the cows.'

'Oh, bottoms!'

I giggled. I had just about got used to the awful swear words my relatives used in everyday speech, but the memory of Madge's 'Bottoms!' still makes me smile.

'Come on, lass. We'll have to start getting the food ready. Thank goodness Annie Lloyd will be here tomorrow to give us a hand.'

Annie was a great friend who walked up from the village twice a week or more. She was a single mum whose daughter, to Annie's great credit and satisfaction, grew up to be head teacher of the local comprehensive school. Annie helped with the cleaning and cooking and with mysteries like pig killing and its products. She also gave advice on all sorts of farm problems and the ways of the locals. A certain amount of gossip also came our way from Annie.

Madge stood on a chair to take down one of the huge lumps of bacon hanging on meat hooks in the ceiling.

'Put one of the big milk bowls on the slab in the dairy and bring some water to tip over the bacon.'

The sides and hams of the pigs were cured in layers of salt. They really needed two days of soaking with several changes of water to clear the worst of the salt from the flesh and make it fit to eat, but one day was going to have to do. I ran down to the barn then and fetched a bucket of potatoes and a bucket of swedes.

We peeled a great quantity of potatoes and put them in a bowl of salted water. A dozen or more swedes were prepared, which Madge decided to cook and mash there and then so they would only have to be reheated next day. I cut up two huge white cow cabbages while Madge made a parsley sauce. This went over the cabbage, when it was cooked, so that also only needed reheating. I peeled and mashed and chopped and stirred, amazed at these mountains of food and Madge's cool organisation of it all.

At lunch (I had learned that we ate lunch at midday, not dinner, as in my old life) Madge asked how many men were likely to be coming.

'Will there really be enough left over from the funeral to run the machines?'

'Some is coming from farther away,' said David. 'No trouble. Not when there's newcomers to see over.'

He banged the salt cellar down and shook it. We three froze, waiting with our usual fascinated horror for him to blow down the hole.

'Hmm. They'll want to see what sort of mess we're making of running the place, I suppose,' said Cecil.

David grinned and said something in Welsh to Tom, who grinned and looked sideways at me. I decided it would not be long before I learnt Welsh.

Cecil saw the look too.

'And there's a new young filly to look over, as well, of course.'

I stuck my nose in the air. I reckoned I could prance and toss my head with the best of them.

In the afternoon I helped Madge put together a great mix of ingredients to make two huge bread puddings. Stirring this mixture needed as much strength as any farm job. These puddings were put to cook immediately in the oven beside the fire. Although it was an unreliable beast and difficult to organise, its slow, unsteady heat suited the undemanding bread pudding well enough. It cooked perfectly in time for lunch next day.

Next morning we were up and finished milking and breakfast in good time – by no means a regular state of affairs – and I went out to listen for the hooter. The first blast of this was blown by Johnnie Thresher as he left home two miles away and it was repeated several times along the way. If the weather was too bad for us to hear it the first time, there would be several more opportunities before it arrived.

Johnnie Thresher's tractor ran on TVO – a form of paraffin. It was an E27N Fordson Major, commonly known as the Galloping Major. It was first produced in 1943, superseding the Standard Fordson. This Galloping Major, pulling a three-ton threshing drum, filled the road from hedge to hedge. It took 30 minutes or more to trundle its way the two miles through twisty narrow roads and down a formidable hill to reach us. All along the way, every 10 minutes or so, the hooter hooted to call the chosen to work. The hooter worked off the same air intake system as the tractor engine, so every time it blasted off, the tractor was starved of its air requirement and the engine died away. Each time this happened, and just before the tractor failed completely, it was expertly resuscitated by Johnnie.

The men in the threshing gang, and they always were males, would leave home when they heard the hooter and either go all the way across the fields on foot or else get to the road and jump onto the tractor and drum as they passed. They hung precariously on to any hold available to get a lift down. An extra long blast when the tractor arrived on the yard and another when they were ready to start, allowed men living close by to make a last-minute dash.

When I heard the blast sound really close, I shouted to Madge, 'There's the hooter' and she set about organising the first mass tea supply of the day.

The man above the ladder is just cutting the band on a sheaf to feed it into the drum. A rickety looking loader is conveying straw from the other end up into the barn where it is being spread to make a tidy stack.

'Come and give me a hand when the drum turns up.'

Back in the stackyard, there were already three or four men standing about among the tall round stacks, talking loudly and vigorously in Welsh. When they saw me they stopped. They all eyed me with great interest, then burst into what were obviously comments to each other and questions to David. There was a roar of laughter and they all turned away, not really embarrassed but not wanting to seem rude.

I grinned brightly, very uncertain of myself. I wanted to go back the way I'd come but instead walked forward and said, 'Bore da' (Good morning) to them and asked David how long before the drum arrived. He answered in Welsh, but then said, 'Half an hour.'

As I walked away, someone called out something and I looked back. They were all laughing and muttering together. Tom was pushed forward.

'Come here, Sheila,' David called. 'Tom wants a kiss.'

I had only recently learnt about kissing, but I wasn't going to let them know that. I was encouraged by the fact that Tom didn't look too eager himself.

'He can come here then.'

Tom was blushing and struggling to get away, so I decided I could risk a bit of bravado and I took a step towards him. He broke free and ran away through the barn. The men gave a shout and chased after him. I wandered along behind, but not so close that I couldn't make a more dignified escape than Tom if I needed to.

Several of the older men had stopped in the barn to nose about. They checked the corn to see how clean and dry it was or wasn't; looked at machinery to see how well it was cared for, seeing what was new, counting costs. They checked around for things that might have gone wrong, like blocked drains and rat damage. Anything to make a story to tell at their next threshing.

There was no sign of Tom.

The machines were roaring and rumbling down the lane now. I ran in and told Madge, 'They're coming!'

We carried out two great cans of tea and served mugs all round. Madge left me to look after seconds and to take tea to the men setting up the threshing machinery. As I took a mug across to Johnnie, a great blast from the hooter went off right by my ear, making me jump and spill the tea. From the bellows of laughter all round, I reckoned it was done just for that effect. I grinned and said some of my newly acquired vulgar words at them. Entertainment was very basic and my ignorance and pushy determination to have a go at everything provided many an opportunity for jokes and japes.

Johnnie was very skilful with his machines, keeping them in good working order and manipulating them from place to place. Setting-up for threshing was a work of art in itself. The heavy, cumbersome box had to be unhitched and manhandled round into just the right place for the job. Johnnie nosed it gently with the tractor radiator as other men wedged it in place. Once in its working position, it had to be made level and firm. Johnnie would hack out different sized holes for the wheels to be lowered into, jacking up one or more until everything was just right. If not level, the machine's sorting system would go awry, mixing up weeds and seconds with top-grade corn.

The power to drive the threshing drum went from the tractor engine to the drum through a hefty belt and pulley system sticking out on one side. The turning of this belt exerted tremendous forward and sideways pull on both machines, requiring a deal of wedging to be employed to control the juddering and leaping about and ensure the belt didn't go slack. The tool for this was a thick plank laid between the front and back wheels on one side of the tractor. The plank sloped off the top of the front wheel and down to wedge under the bottom of the back wheel, perfectly counteracting the pull.

While all this was going on, the rest of the men were opening up stacks and generally sorting out the work to be done. There were important and unspecified rules about cooperative working which you learnt mainly by watching and chatting and joining in. For example, a newcomer did not try to tell workers what to do. Certain jobs went to certain men, and although someone might come to the owner of the farm and ask, 'What do you want me/us to do?', this was a politeness only. The men knew the routine and their place in it and they just quietly got on with things.

Two or three of the older men worked in the dutch barn, taking in the straw and building it into a stack. A lot of knowledge and skill went into making a tidy stack that wouldn't fall down, but still it was fairly light work, leaving time and breath for a rest and a clonc (chat). The loose straw was passed onto the stacks by an elevator. Where this could not be used because of the shape or position of a barn, the straw was gathered and passed up on a pitchfork, which took skill to do well and was slow.

Another light job done by the older men was watching the four sacks filling at the grain spouts on the back of the drum. As one became full, a slide was pushed across to stop the flow of corn. The sack was moved away and an empty one clamped in its place. Two of the spouts delivered top-quality grain, another was seconds, the fourth was rubbish and weed seeds. The strongest men had the job of carrying the heavy sacks of best grain across to the barn, up narrow, almost vertical steps, to empty them into the granary. These sacks filled every few minutes, so the men carrying them were kept very busy.

Threshing. Freestanding stacks are taken apart and the sheaves passed to men on the drum. Sacks are seen filling with corn. The threshing machinery is powered from the tractor through the belt. The tractor is Johnnie Thresher's Fordson Major, E27N.

A young boy usually got the unpopular job of clearing the cavings – the outer husks of the grain – which were blown out at the other end of the drum. Wandering around eager to help, I got myself landed with this job, getting many a sidelong, grinning glance in the process. I very soon discovered why everybody found it so amusing. It was dusty and very itchy. Dust and bits got everywhere. The worst were the barley whiskers, and there was almost always barley among the corn. These sharp, hairy pieces of dried plant got in your socks, down your neck, up your sleeves. They worked their way into every strand of clothing and every bodily crevice. And there they stuck. Not just for that day but for many another. Even after washing, clothes and body remained scratchy and scratched.

By the next winter I was a good enough worker to join in with pitching sheaves from the stacks to the drum and working on the drum itself, as efficient as any of them. I enjoyed the rhythm of the work, especially catching the sheaves as they were pitched to the drum, cutting the strings and fanning the heads out to feed them into the threshing machinery, two of us alternately catching and feeding, without stopping. Pitchers and feeders worked with a graceful lift and swing in their movements. They kept to the same rhythm so that no one was kept waiting or got overloaded with sheaves. Occasionally there would be a hitch. Perhaps sheaves would bind together in the stack or be too wet or mouldy to use. Or a feeder might drop his knife, which should have been tied round his or her wrist, into the works.

There were enough men at our February threshing for two sittings at lunch, and thanks to Madge's organisation, everything went smoothly. Dishes were shuffled between the paraffin stove, the fire oven, and a precarious primus stove. I don't know how it was done, but with the help of Annie, plates of meat, dishes of vegetables, jugs of parsley sauce, bread pudding and custard, were all served piping hot to two sittings of men. And when they had all gone back to work, Madge, Annie and I sat and had ours.

Another set of workers came with the men. These were the dogs: little terriers and corgis and mixtures of the two, who were expert ratters. They chased and killed the rats displaced from their cosy homes in the stacks, often three or four dozen of them. Few escaped.

There was great satisfaction in seeing the bright, golden grains of corn piling up, knowing we had food for the beasts for the rest of the winter and seed for next year. It was the culmination of all the work and worry of cultivation, sowing and harvesting.

This year's crop was not one of the best. The wet autumn meant sheaves never dried properly and eventually had to be carried and stacked damp. There was mould in the grain and straw, which spoiled the food value. The grain had to be constantly shovelled about during the winter, shaken and turned to get air in to dry it as much as possible. Harvests were often like this in Wales. Many years later, farmers bought electric fans to blow air continuously through the grain. This did a much better job. Later, with combine-harvesters gathering and threshing the grain so quickly, it was rarely damp anyway.

Chapter 6

No Snow to Speak Of

THE real difference between the winter of 1947 and all subsequent bad snow years is not measured in quantities of snow, strength of wind or lowness of temperature. These were probably all much of a muchness in '47, '63, '65, '82, and whenever. The 1947 snow-up was outstandingly memorable because it was the last one before modern technology made clearing up easier and survival harder.

When the big snow of 1982 clogged up the working systems of Ceredigion, the local paper described the weather as the worst of the century. Another said the worst since 1947. Certainly I remember the blizzard conditions lasting longer in '47, and the snow being deeper in the roadways.

But as in all situations, the facts that matter and stay in the mind are those that affect us personally. For my husband and me, 1982 was not the horrendous experience it was for many of our friends and neighbours because we had given up producing milk. We didn't have to battle with frozen milking equipment; we didn't have the worry of storing and throwing away hundreds of gallons of milk; and we didn't have the desperate need to dig and drive our way across three impassable miles to the main road to meet the milk lorry when it was able to get there. I think anyone who was dairy-farming in the district in both 1947 and 1982 will remember 1982 as by far the worse time, not for the amount of snow that fell but for there being vastly more work, wastage and stress. With five times the number of cows, 10 times the amount of milk, thousands of pounds invested in valuable machinery and overdraft, there was a lot to worry about. An utterly irrelevant but interesting point about 1982 was that the snowfall came at the beginning of January. All the other big falls that I can remember came in the first week of March or the last of February.

The very first snow of 1947 fell right at the end of January, but this was only an inch or two. Madge had gone up to London to receive her retirement present from the Civil Service and she was due home in the first days of February. A further fall of snow with sharp frosts had made roads treacherous. I don't remember if buses were getting through to the village, but anyway Cecil took the car to Newcastle Emlyn to meet her. The drive was slithery and even more bumpy than usual, but he got there and back safely. There continued to be small falls of snow over the next three weeks and with sharp frosts at night, the ground was very cold and hard. This made moving about difficult for man and beast.

Serious snow began on 4 March. The first flakes came drifting gently down at about three o' clock in the afternoon, big and slow and beautiful as a Hollywood romantic movie. Within the hour there was a full-scale blizzard.

I decided I would do an Arctic Expedition into it.

I set off across fields I hardly knew, tremendously excited by the swirling of the storm around me. I was probably making up some wild adventure story about getting through prodigious hazards and hardships to perform some desperate and noble deed. I mostly lived my life in such fantasies.

But fantasy blew away when I turned to go back.

I was now facing into the north wind and it was blowing thick, blinding snow horizontally

Looking out the morning after the blizzard, 6 March 1947.

Cowshed with 'lace curtains'. The snow filling the lane in the distance rises higher than many of the trees and bushes growing on the 5ft wall.

against me. Strands of hair froze into bundles of icicles that whipped viciously at my face and eyes. Snow piled up inside my loose-fitting woollen hood like a fluffy lining. I couldn't open my eyes, but had to squint sideways between my fingers. Not that seeing helped me. The landscape was already unrecognisable. I did at least know the direction I'd come from and as hedges were discernible by the drifts building against them, I had only to follow this line and I must arrive home.

I reached what had been a corner. Tops of trees in the hedges indicated the turn and I found myself slightly sheltered as I changed direction. I knew that not far away now there would be a gate. But I found no gate. It must have been farther along than I'd remembered.

I struggled on but still no gate appeared. I began to panic. If the gateway had filled up, I would be trapped and buried in the snow. I would not be able to climb over the soft drifts and no one would find me. No one knew where I was. It would be an impossible job for them to search the whole farm for me in this weather and any shouting I could do before being completely buried would be gulped down by the storm. Head down, I pressed on.

Then suddenly, incredibly, there was grass under my feet. Shading my eyes and flicking aside the ice-bead curtain, I saw a clear swept track that stretched out of sight to my left between two vast snow banks. Here was my gate. Luckily it had been left open, which allowed the wind and snow to rush straight through and away across the field until it was stopped again at the next hedge. One of the things that astonished me later was the amount of empty space that was left between snowdrifts and the way centres of fields were swept so entirely clear of snow.

I arrived safely home and no one had even missed me.

Carrying water to the cows that night was a great struggle. I suspect they didn't get quite as much as usual.

Next morning the blizzard was still raging. We had to dig our way through snowdrifts to get out of the front door and then battle through the snow and wind to reach stable and cowsheds. Animals had to be fed and watered, cleaned and milked. The pigs were quite cosy in their substantial stone sties. Snow had piled up in a windbreak outside their front doors between house and run, but they got out to their troughs without any trouble. I don't remember what was done for the chickens, ducks and geese. The chickens must have had food and water put inside their house, for they would never have ventured into a blizzard. Ducks and geese must have been let out briefly under guard to gobble what they could. Even the geese would surely not have attempted to wander off.

Some time during the night, after 36 hours of unabated raging, the blizzard stopped as suddenly as it had begun.

We got up to find a solid wall of snow, five and a half feet high, once more blocking our exit. There is a photograph of the porch with me and Cecil behind this wall, our faces just visible. The mystery is, how did Madge get out to take the photograph? There is no sign of the wall of snow having been violated in any way and I'm sure the back door had not been installed at that time.

The rearrangement of our landscape was impressive. On the north side of the yard, the house and cow-house were almost obscured by great mounds of snow up to the eaves. The drifts stood away from the buildings leaving a walk-way between them going from house to cow-house. On the other side of the yard there was barely an inch of snow lying.

A little farther down, the shape of the yard and buildings had drawn the snow close in to the cowshed where it built up to a height of 10 feet or more. A cascade of icicles hung from the roof. Heat from the cows inside had continually melted the snow on the roof and during the night it had dripped and frozen till it made a sheet of frosted glass stretching from the eaves to the ground.

Up the yard beyond the buildings, the snow could be seen packed into a great causeway. Sloping gradually upwards it filled the lane and spread over the hedges. When we later had to walk out for necessities like paraffin, and luxuries like recharging accumulators to play our wireless, the tops of hedges were our highways. In many places you were high enough to touch what were then called the telegraph wires. These treks, about four miles the round trip for accumulators, five or six for paraffin, we did several times before the roads were cleared.

Lack of modern technology meant there were few tractors and none of them had any kind of snow clearing equipment. Even the War-Ag. – a wartime branch of the Ministry of Agriculture and Fisheries which hired out machinery and told farmers what crops they were to grow, often utterly unsuitably! – had nothing that would make an impression on this snow. Efficient and specialised fittings like scoops and snowploughs came much later, after the installation of hydraulic systems on tractors.

No, this snow clearing was done by hand with picks and shovels; and a lot of those hands were supplied by German prisoners-of-war from local camps, and by Polish soldiers from a camp at Aberaeron. While farmers and other able-bodied citizens worked on their own personal access areas and the minor roads, these POWs were digging their way steadily along the main roads to open up access between towns.

I don't remember precisely when David made his way through to help us, but it was not many days after the fall. I do remember we were astonished and very pleased to see him. I think Tom came down the first day after the snow stopped.

I helped feed and water the cattle. There were no self-feed arrangements. No baled hay either. Sections had to be chopped out from the haystack. This was hard, heavy work and the big, curved knife – shaped rather like the French guillotine blade – needed frequent sharpening. David was the only one among us who could do the job efficiently. The cut hay was put on an opened-out sack whose corners were then gathered together and the bundle carried across the yard to the animal houses. Packing the maximum amount on the sack so you could hold the corners together and carry it was a skilled business. You had a big and cumbersome load, very susceptible to windage. Straw was carried for bedding in the same way. These were woven sacks, of course, not paper or plastic.

Water had to be carried by the bucketful from the well across the yard. It was then held up to about chest height while each cow reached over the partition to drink. They did not always get as much as they would have liked, especially those not in milk. Ten gallons each a day for 12 cows made an awful lot of carrying. Once the yard had been well enough cleared of snow and ice, they were taken out once or twice a day to drink their fill at the pond. This saved a lot of carrying and made cleaning and feeding easier. We took only one or two cows at the same time for fear of them slipping. A herd of cows is very like a crowd of children: they push and shove, all wanting to be first, but the dominant ones push and shove hardest and there are casualties. It was the custom locally to tie cows in their sheds for the entire winter, usually through October into April. This was to save the land from 'poaching', that is tramping it into mud in the wet weather. Poaching killed or greatly inhibited growth of spring grass. Cecil decided it was better to let them out for an hour or two, even all day in good weather, for air and exercise, keeping them to the one transit field.

There were always worries that stores of cattle food would not last the winter. The worst years produced barely enough hay, corn, straw and roots to feed the animals through an average length winter until the spring grass came. A bad summer, like this past one of 1946, meant not only a poor corn harvest but also an acute shortage of grass through all the growing season. By the end of a year of bad weather like this, cows could be quite hungry; as indeed they would also be in a drought. I have climbed trees to break off branches for cows to eat the leaves from in drought conditions. The conditions

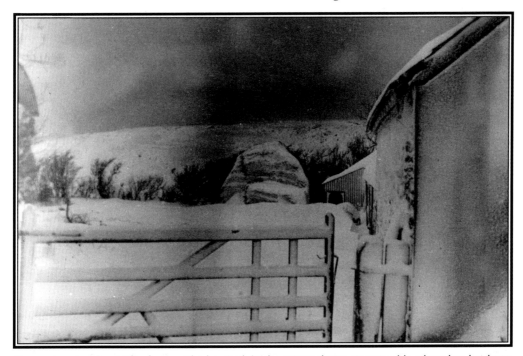

Haggard under about 5ft of snow. The haystack in the centre shows an open side where hay has been cut for feed.

Looking down the yard, 1947. The bush in the foreground is a huge butcher's broom which went absolutely mad, left to itself.

would be alleviated by feeding hay, straw and corn, but all of this food had to last through the winter as well. Extra could be bought in if you had the money, but hungry cows made less milk, and less milk meant less money. You always hoped desperately for an early, warm spring to give a bite of grass. In this exceptionally long winter, when storage sheds contained short rations of poor quality food, we hoped in vain. Snow still half-covered the fields well into April.

There was no worry about going short of food for any of us humans. For the likes of us with a farm, there was always an abundance of food to survive on. We didn't have the foods that nowadays would be always in store cupboards, like tinned fruit, vegetables, fish and meat; like sugar, flour, dried fruits, tea and coffee. All of these were tightly rationed goods and supplied in small quantities by the week. But still we ate well, as always.

We had tons of potatoes and swedes, plenty of apples, home-cured bacon, and, of course, milk and eggs. We also made butter. We ate the occasional chicken, usually a sickly one before it died. And we shot rabbits. This was luxurious living by any standard at that time. Indeed, I would say it was so for any time.

Chapter 7

Surviving

THINK what panic and complaint there is when gales or snowfalls deprive us of our phones and electricity, even for a few hours. Especially electricity, which has so many essential tools of everyday survival dependent upon it. It was not so great a loss in 1947, for those who had mains electricity had fewer gadgets than are about today. There were none of those desperate freezer crises with all that costly food going to waste, or possibly poisoning us! Also the war had accustomed people to coping with losses and shortages.

The national grid system of electricity was set up around the British Isles in the thirties and it even reached into the far west of Wales. A power station in Pembroke Dock supplied towns along its route northwards, going through Pembrokeshire and Cardiganshire. Sometimes along the way, as happened with railways as well, there were rich and important people a mile or two to one side of the route. For these special arrangements had to be made. In the case of railways, this would be a private station. With electricity it was a spur taken off the main line to supply power to a big estate. One such spur supplied the Liberal MP for Cardiganshire, Sir David Owen Evans, who lived in the old mansion of Pigeon's Ford about two miles away as the wires stretch and one mile inland from the village of Llangrannog. He kindly allowed the village to tap into his supply, but the people had to raise money themselves to pay for the installation.

These arrangements did not stretch to isolated holdings and so we had no electricity to lose or miss. Milking could go on as usual; light was available as usual; heat for comfort and cooking was there, as usual. Farms and other isolated dwellings were already accustomed to stocking up with plenty of non-perishable food to take them through bad winters, so eating was much as usual too. In fact the whole of life for us at Penrallt went on pretty much as usual.

Villages were not so well stocked as farms. When supplies in homes and shops ran out, little fishing boats put to sea to fetch bread from bigger towns with bakeries. From Llangrannog the boats went to New Quay, five or six miles up the coast. With no convenient plastic materials for wrapping and packing, bread mostly arrived soaked with seawater. Once or twice the whole lot was lost overboard when an extra big wave hit the boat.

On the first morning after the snow stopped, Madge and I set forth with shovels and food and water to let the fowls out. We dug our way through to the stack yard, then made paths and feeding areas in front of the duck houses. We cleared some of the snow off the little duck pond and broke a hole in the ice for them to swim and drink. When we opened their doors, it was clear that these creatures, too, had experienced no snow to speak of.

They didn't notice anything unusual at first, but rushed out with great enthusiasm after their long confinement. In their eagerness they got wedged together in the narrow doorway, tripping each other up and tying themselves in knots with their great feet and long necks. Once outside they paused, staring about with their big round eyes and Buster Keaton faces. Then with a frantic flapping and slipping and quacking, off they went down the icy gully and launched themselves onto the pond. It was a sight Walt Disney must have had experience of when he drew his duck cartoons. They did the splits, they skidded on their long necks or on their sides across this strange surface. They stood up and fell down in a riotous circus performance. Their big feet and wide, bewildered-looking eyes were perfect clown make-up. It took them quite a while to find the hole we'd broken for them in the ice, but once they were in, the warmth and flapping and squabbling of half a dozen duck bodies soon made it considerably bigger. The ice was not thick. A few days later it became too solid to break on this shallow pond. The ducks were not at all eager to get out again, once in, so we left them to it while we went to see to the hens.

The hens were much more circumspect. They crowded in their doorway, pushing and pecking to get a place, jumping on each other's backs, poking out their sharp little faces to look this way and that. One would tentatively extend a claw to test the snowy ramp, draw it back and retire. Then another did the same. We clucked at them and rattled the corn in the pail, but none would risk coming out. A busy chattering went on among them the whole while, with now and again a certain amount of louder cursing and complaint as they got in each other's way. Eventually a couple edged like tightrope walkers down the ramp and scurried to the food trough. Their pecking excited the others who started to come out in wild dashes, one and two at a time. Some launched themselves into flight to avoid the strange substance on the ramp, crashing past others still hovering in the doorway. Suddenly a terrific squawking started up inside and one bird came hurtling from far back, flapped over the top of them all and landed in a flurry of soft snow. She stood shaking her feet and shuffling her feathers tidy, just like any respectable matron caught off dignity. We gurgled with laughter at the bewildered expression on her face. And anyone who thinks a chicken incapable of feeling or showing bewilderment or dignity is not acquainted with your old-fashioned, independent free-range bird.

Only the geese pretended they'd seen it all before, stepping out in their usual disdainful fashion. Gander swung his neck protectively over his women and hissed aggressively as he hustled them into the open field. They went mountaineering over the slopes, sometimes sitting back on their tails to slide down the steep sides. They made their way quite easily to the big pond which had a large area unfrozen where the spring flowed in. When it came to getting them shut in again at night though, they had quite forgotten the way and it took an annoying age to herd them back to their shed. The ducks were equally troublesome, not wanting to be shut in again. Only the hens just

Snow woman in red and blue feather hat.

Sheila returning from a five-mile 'walk' with two gallons of paraffin.

went quietly home and sat on their perches, waiting patiently for the next day and a new lot of food.

The biggest problem we had, in the three weeks when we were cut off from the rest of the world, was what to do with our milk. Cows could not be switched off for the duration, nor could the production unit be closed down. We wouldn't have been getting more than six to eight gallons a day at most (27 to 36 litres) – a miserable pittance by today's standards – still, that constituted our entire income. At the grand price of about one shilling per gallon, it was not to be lightly tipped away! The wonderful Milk Marketing Board, sadly now closed down, promised to take and pay for at standard rate, all milk produced whatever its condition. So all we had to do was store it for as long as it took.

The system in everyday milking was to put your full churns on the milkstand and receive an equal number of empties in return. If you were sending three churns a day and then the amount of milk dropped to fill only two churns, you still had three empties to take back from the day before. You were supposed to send back the spare one but most people kept a few in reserve for when production went up again. Also churns were very useful for transporting water to fill cows' drinking-water tanks in distant fields. We probably had three or four spare. Each churn held 10 gallons, but they were never filled to this level as the milk slopped out with every bump in the road. For storage, filled to the brim, they would take maybe four to five days' supply. After that everyone filled every bowl, bucket, saucepan, wash basin and jug, every possible or impossible utensil they could lay their hands on including chamber pots, those who had them.

The house dairy, the long narrow extension beyond the kitchen, had a concrete floor and thick slab slate surfaces for cool storage. These slabs were about 30 inches high and ran along each side of the room, supported on brick pillars. They were maybe 24 inches wide and seven feet long. In normal times two or three bowls of milk stood on these slabs, but in the snow as many containers as possible were packed there. We consumed quantities of milk in all kinds of dishes, and we made stacks of butter. Some of our milk and butter would have gone to friends and neighbours and others in need in the village. We even tried our hands at making primitive, mostly disgusting, cheese. Pigs, dogs and cats were always glad to help out, but in the end, some milk had to be thrown away.

Again, there was less waste than people today might expect simply because the herds of cows were so small by modern standards. The number of cows kept on this farm of 80 acres – and other farms were similarly stocked – was only a dozen. Twenty years later there would be 30 on the same acreage and a few years after that, 60 and more. It was not just the numbers of cows, though, that made the difference. There was also a great increase in the amount of milk produced by each animal. They were bred and fed, like crops and the soil, for maximum production. Whereas cows in our first 10 years of farming produced on average 400 to 500 gallons (1,800 to 2,250 litres) and were reckoned very good milkers if they produced 600 gallons (2,700 litres) in a year's lactation, by around 1980, production of 800 to 1,000 gallons was normal. This has continued to rise, and surely such extreme pressure on animals must have contributed to the great increase in old and new diseases afflicting them. And it certainly increased storage problems for farmers in later heavy snowfalls. Even with the roads getting cleared so much more quickly, there was still far more milk thrown away than in the sparse and spartan days of 1947.

Back then, every able-bodied person was digging snow. Three weeks after the big fall they were still digging along the minor roads. On the main roads though, the organised gangs of German POWs had dug their way through the 12-foot drifts and made a single track wide enough for a lorry or bus to drive along all the way from Cardigan to Aberaeron and inland to Newcastle Emlyn and Carmarthen. Buses could now run between towns and railway stations; and milk lorries could pick up milk from appointed places on main roads.

After three weeks we heard, at long last, that the milk lorry could get through on the main road and would be picking up churns next day at Brynhoffnant. All we had to do now was to get them there. The little sledge Madge and I had slid down the hill on seemed the only possible thing we could use. Four churns were tied very precariously onto it, and David, Cecil and I set off across the fields on the two-mile drag, incorporating a 200-foot rise to the pick-up point at Brynhoffnant.

Quite soon one churn fell off, spilling most of the milk, and it was obvious that the load was too heavy anyway. We stood another churn aside and struggled on with two. They were still difficult to keep stable and very hard work to pull over the rough ground. We took it in turns, but still David ended up doing most of the heavy pulling. A dozen or more men were already at the pick-up point when we got there. They came from surrounding farms, some with sledges, some with wheelbarrows. Those with access to the main road got a horse and cart along. Few could have had such a hard and hazardous trek as we did. I don't remember how long it took us to get there. It must have been at least three hours. It was more than half-an-hour's fairly steady plod up the road in normal conditions. I do know that by the time the lorry had come and we'd given in our churns, collected our empties and got home again, a whole long, cold day had gone by.

Almost as important as getting the churns away was getting empties back. They were supposed to be rationed out, one empty churn for one full, but quite often people getting there first would take more than their fair share, leaving others without. Farmers living near pick-up points were lucky all round.

Side roads remained impassable for another one to three weeks, depending on how important they were. Bus routes to villages had priority and the road to Llangrannog was one of the first to be cleared.

It's strange to remember that snow lay on the beaches right down to high-tide mark for a long time. At higher levels it gathered in caves and corners of cliffs and among the sand dunes. Little bits lingered as late as May. In sheltered corners of high inland hills, strips and mounds of a dirty icy snow were still about in June!

My main memory of those weeks is of endless, brilliant sunshine and clear blue skies. There must have been some grey days – I do recall a few – but mostly it was so hot we were sunbathing. There were huge amphitheatres of glistening snow which had been beaten out by the wind. They were swept up into fragile curves like the tops of waves about to break. We lay in these with the hot sun beating down, reflecting off the surfaces, and a cloudless blue sky overhead.

It was during this time I first saw the North Wales mountains across the great sweep of Cardigan Bay. The sea and sky were bright blue, the ring of mountains clear and white. It was a sight that has never lost its excitement and glory for me. On a clear, grey day, a slightly hazy day, or in the brilliant gold and blue of perfect summer, there is always something magical about the sudden materialisation of those humped mountains edging the sea.

But never again were they so breathtakingly beautiful as in that first snow when they flashed their frosty diamonds against the sky and the sea sparkled with sapphires.

Village men carrying a boat down to the sea to sail to New Quay for provisions, mainly bread.

Chapter 8
Pig Killing

I F YOU are a vegetarian, or just plain squeamish about the business of turning animals into meat (in which case I assume you avoid those ferocious programmes on TV where wild animals hunt and tear each other to bits for this purpose), then you may decide to skip this chapter.

Two of Madge's closest friends were vegetarians. One of them used to say that the only time she felt really tempted to eat meat was when she smelt one of our hams cooking. Besides being vegetarians, they were also atheists. They had a daughter a little older than me, heartily self-assured and independent, whose easy attitude to life and people filled me with an amazed awe. They brought with them on one of their visits the playing cards and rules for the game of canasta, which had just become all the rage. My memory of the rules is very hazy now but I do remember the tremendous fun we had playing the game. More than anything, I remember the astonishing amount of personal verbal abuse that went on, particularly between Stella and her father. I can still feel my eyes widen, my mouth open and a great laugh rise in me when I think of her turning on him in fury after he won the huge, tottering pile of discarded cards, saying, 'You dirty great heap of steaming horse-shit!' Stella worked in stables and no doubt used this word in its basic sense all the time, but it was certainly not a word normally heard during polite parlour games. Even Cecil used it only occasionally and then only in reference to the substance itself.

Sheila Barefoot with the breeding sow, Belinda. Looking roughly north across the Home Field towards Cefn Cwrt.

However, from this unlikely nuclear family of vegetarian atheists has grown a dynasty of meat-eating Christians. This could be a fascinating item for those interested in the nature/nurture controversy. Stella took to eating meat quite early in her life, later keeping chickens and killing them herself, as anyone who eats meat should be prepared to do, albeit as kindly as possible and with all due respect. The very nicest way I ever heard of was from a well-known (well, he used to be well known) self-sufficiency man, John Seymour, who farmed in Pembrokeshire and wrote books on the subject. Possibly he still does all these things for I read the other day that he is in his 90th year and his classic *Self-Sufficiency* book is still in print. He said that when he wanted to kill a pig, he shot it while it was eating its food. Presumably he had a hoist, or the equivalent in human hands, to then quickly hang it up to bleed.

All sorts of people kept a pig during the war. Even in towns they might be found at the bottoms of gardens or on allotments. They were probably killed by the local butcher. Food was whatever kitchen scraps were available, augmented with left-overs from friends and neighbours, though left-overs of any kind tended to be meagre and of low nutritional value in the war. Some of the pig meat would be shared, but probably not a lot because one year's ration of bacon was given up for the privilege of keeping a pig. I wonder how good a bargain this was, for they mightn't have grown very big on just waste food.

At Penrallt we kept and killed two pigs a year, so two ration books had their bacon coupons cancelled. There is no question of it not being a good bargain for us. Our pigs were abundantly well fed. Kitchen waste was plentiful and good, often including skimmed milk. There was also a sort of pigs' porridge made from ground barley. Our pigs grew very big indeed. Too big, we soon decided, even for our acquired taste for very fat bacon. The standard size was 20 score. That is 400lb or 3.5cwt. Or 180kg in new money. That is an awful lot of pig. The streaky bacon that came off its sides was almost solid fat, the flesh part showing as two or three very thin pinkish lines.

We had to be very dedicated to the spirit of the country idyll to actually enjoy eating this.

The killing had to be done in cold weather to avoid flies and to keep the meat fresh during the few days when it had to hang in the open. The first pig would go in November, possibly early December, but not later than the first week, bearing in mind Christmas and poultry preparation were coming. If you had only one pig you might kill it soon after Christmas, but with two you'd want to leave it as late as possible, which could be early March. My first experience of pig killing must have been in February 1947, before the March snow-up. By the time the snow cleared enough for people to get about to do the job, the weather would have been too warm.

Preparations started two or three days beforehand. First of all the great cauldron had to be cleaned out and set up on an iron tripod. This allowed a fire to be built underneath it. The cauldron was three-quarters filled with water and a quantity of fuel was put ready for boiling it up.

Griffy Jones, who had the farm before us, was our local pig killer. I don't know how long he continued with that job after he gave up the farm, but he was certainly doing it in our first year there. I only vaguely remember this but when I was checking other facts recently with Dai, who lived on the farm next door to us, he was quite certain that Griffy had been killing pigs for people in 1947 and therefore would have done it for us then. He was only a little boy at the time but he was sure of the year because that was when he and his family moved into Nant y Bach, just in time for the big snow. And he remembered it was Griffy, he said, because Griffy's nose kept running and Dai was fascinated to watch the drips dropping off the end and mixing with the hot water on the dead pig as Griffy scraped its bristles off.

Several neighbours came to help with the job, for there was a lot of heavy lifting to be done. Griffy was undoubtedly paid a fee, but the others were probably paid in reciprocal labour and with some of the pig meat. This was not one of the big communal jobs but neighbours helped each other.

The most troublesome part of the job was getting the pig to walk up the yard. Pigs are not natural cooperators. They are strong-minded, intelligent individuals, difficult and dangerous to coerce. Even in the most everyday situations, often leading to their ultimate advantage, they are pig-headed. You have only to mention to a pig, in the nicest possible way, that it might like to move to another, cleaner, cosier home, and it is screaming blue murder, running round in circles and foaming at the mouth, snapping its jaws with a force that would take half your leg off, given the chance.

On one occasion we were setting out to move Belinda away from her piglets. I made up the most delicious smelling bucket of gruel, warm and milky enough to entice any normal pig almost anywhere. It was not enough to make Belinda forsake her piglets though. We got her out of the shed on the chosen day, closed the door and persuaded her, with much growling on her part, halfway down the yard before she fully realised what was happening. Then she swung round and charged back. No amount of whacking, shouting or rattling of buckets stopped her. Somebody tried to get a noose in her mouth to tighten on her nose and upper jaw, which is probably quite painful since it encourages a pig to do what you want, rather like a horse's bit or the ring in a bull's nose, I suppose. Anyway it wasn't possible. We got a rope round her neck, but she was too strong to force along, and too dangerous. An angry pig snarling and foaming at the mouth is one of the most frightening animals to be confronted with and probably as deadly as any jungle beast. She won that round and went back to stay with her young for a little longer.

One similar battle I won though, with a brilliant idea and the coincidence of a pig having been killed a few days earlier. On one of those days, Annie was up doing all the expert, messy business of salvaging anything and everything that was edible from the pig's body. I was taking one of the sows home from a few hours rootling and romping in the field and I passed the out-kitchen just as Annie opened the door and put out a bucket of bones and waste gristly bits. Before I realised what was happening, the sow swung aside from her sedate trot home, stuck her snout in the bucket and champed and slurped with immense gusto at her relative's remains. When I finally persuaded her away, she went on down the yard with a great leg bone in her mouth, scrunching it up like a dog-biscuit. So during the aforementioned pig move, I left the battle we were having with whichever pig it was and I got a bone and tied it on a string. I then dragged the bone down the yard in front of Pig's nose and she followed me all the way without a single stop or glance back!

Hearing a pig out in the open that is being dragged or driven, you would certainly think, if you knew no better, that some poor creature was being most horribly tortured. The screaming was the same if you were trying to get Pig up the yard to visit the boar as it was going to the slaughter table. People often attribute to animals the ability to know they are about to die. This could well be true in the noise and smells of a slaughterhouse. There may even be direct communication between animals. On their home yard, however, in ordinary everyday circumstances, I suspect there is no more intuition of death among the general mass of animals than there is among humans walking out of their front doors to die under a bus or by a murderer's knife. There was no way this intelligent animal would have had any dread that it was on its way to have its throat cut. Its screams were just another gesture of independence. The business sounds barbaric, but the distress, discomfort and pain involved were nothing compared to the trauma of modern day transportation and slaughterhouse death.

So, everything being ready, Claud Pig said his last, mistaken, 'No – after you, Cecily' and began his noisy, troublesome trek up the yard, though getting slightly less his own way as the rope was tight over his upper jaw.

The slaughter table was made of thick, heavy planks, white from years of scrubbing. It was quite low down – 18 to 24 inches high at most – to allow the pig to be tipped sideways then held on its back with one man hanging on to each back leg, another pulling its head back with the rope. The routine was well practised and very quick. The butcher slipped his sharp knife into the jugular before Pig could have had more than the briefest of thoughts about whether this might be some new kind of mating ritual. There could have been very little pain. I judge this from all the information we get these days from people who suffer severe cuts and say they feel no pain until long afterwards. The body threshed about for several minutes, helping to pump the blood out. This was collected in a bucket.

The next step was to scald and scrape the carcass. Buckets of water were carried from the cauldron and tipped, a little at a time and just off the boil, over the body to soften the bristles. These were then carefully scraped off with a very sharp knife, leaving the skin smooth and clean. When all the scraping and scrubbing was done, buckets of cold water were tipped over the body to cool it. A piece of wood was inserted through the back legs to hold them apart and give a solid support for hanging the animal. Then it was carried down the yard, hoisted up and hooked by this piece of wood to the top rung of a vertical ladder. The belly was slit open and all inner organs pulled out into a tin bath. (Was that the same bath we washed ourselves in, I wonder? It would have been well scrubbed for both purposes. Still, there were probably two.) I can't think if anything was done with the intestines. I remember that Annie used to scrape all the fat off them. That must have been a fiddly enough job. I don't know how the entire cleaning process would have been done to make sausage skins of them. I remember only one occasion when Madge and I had a go at making sausages. It was hilarious, messy and frustrating, trying to force all that mushy minced meat into tiny, slimy tubes.

All country women would have known the skills associated with pig killing. Now all these skills are lost. Annie did all the specialist work for us. As well as round the intestines, there were layers of fat round the belly, kidneys and neck, and all of it was put in a large saucepan and rendered down into beautiful soft white lard. This was clarified and stored in large glazed earthenware jars and was used for pastry and roasting and sometimes in cakes. It kept perfectly fresh for years, covered and stored in the cool dairy.

I don't remember the precise lengths of time involved, nor the ways of making all the side products, but quite soon after the killing, all the prime, fatless flesh of the under belly was cut away and taken into the house. There was no way of preserving this meat so it had to be eaten within 10 days or so, depending on the strength of your stomach. Stomachs were much hardier in those days and learnt to fight back against all sorts of bugs, including those connected with food deterioration. There must have been several pounds of this meat because the custom was to give packs round to neighbours. Then those who had pigs of their own would do the same for you when they killed. It was another of those communal interchanges that worked so well, equivalent to a deep freeze (only more reliable!) because you had the fresh meat to eat at two, three or more later dates instead of having to eat it all at once. In addition to these local exchanges, Madge also sent off little parcels to friends in town. With meat still very meagrely rationed, this was a great treat and much nicer than anything you could get from a butcher's shop. I was always given some to send to my Auntie Kathleen.

While Madge and I were packing parcels, Annie Lloyd was sorting, cutting, mincing and cooking the offals and head to make brawn and faggots. Annie's brawn was a delicacy superior to any paté I've ever tasted. The finely minced meat and the jelly were perfectly balanced and mixed, the flavour full and smooth. Her faggots were likewise splendid, and even allowing for time's wear and tear on memory and taste, I think they were better even than those of the best butchers in Wales, the Williamses of Machynlleth.

Salted bacon in muslin bags, hanging from hooks in the living room. Cloths hang to dry on a wire inside the chimney beam. One kettle is suspended over the cwlm fire, the other sits on the oven top. Above right is the bread oven. Against the other side of the fire is the water boiler.

After three days of hanging, the bulk of the pig was ready for cutting into sections to be salted for bacon. The dairy was cleared and the great slate slabs washed down. A large quantity of salt and saltpetre was put ready and this had to be rubbed into the flesh of the sides, shoulders and hams twice a day for three or four weeks. The sides were finished first. Because of the thinness of the flesh and the absence of bones, the salt penetrated quickly. The large hams and shoulders took a lot of working to make sure a good preservation barrier was in place. Most particularly, salt had to be forced down beside the bone and into its cavities, for those were places where flies might sneak in and lay their eggs.

As the pieces became ready, a sharp-pointed, S-shaped steel hook was pushed through the flesh of each joint with more protective salt put round the holes. Then each joint was wrapped in fine muslin. Edges of the material were sewn up and the top opening was tied very tightly round the metal hook. They were then hung up on the ceiling hooks in the living room and left to mature until needed.

I don't remember when we stopped eating our own pigs. Certainly after February 1953, for I have a letter talking of the second pig being killed then. More stringent laws came in about how and where you could kill animals and probably we didn't want to be bothered any more. Meat and bacon were still rationed as late as 1954, but the quantities were more flexible, certainly in country areas, and there seemed to be a little more offal and fish available. The whole farming system was changing too. Mixed farms were going out of fashion. Specialisation was the name of the game and reckoned the only efficient way to maximise production and profit. Cardiganshire was one of the best milk producing areas in Great Britain, so it was sensible to concentrate all your land and energy on that. There was no time or place for pigs, fowls, horses; nor for the growing and harvesting of crops like potatoes and fodder-beet. There was stronger and more convenient nutrition to be had in sacks.

Chapter 9
Sleeping Out

EARLY on in the year and lasting until unremitting winter set in (usually November!), all bedrooms in the farmhouse were vacated for the benefit of Paying Guests, plus the sheer joy (I was told!), of Sleeping Out. Of all the many outrageously alien concepts I was introduced to at Penrallt, only this business of sleeping out sometimes caused me disloyal twinges of doubt regarding its overall merit. Making room for guests was one thing. Enjoyment, as far as I was concerned, did not come into the business at all. I did try. And on a perfect summer night, with a full moon and a sky brilliant with stars, it was magical to walk out across the field and then just lie and gaze up at all that black and gold and twinkling blue as you went to sleep. But nights like that were fairly rare and it was more the sort of thing I'd choose to do on occasional nights while on holiday.

Certainly I was quite excited the first time we embarked on this exploit. I had never been in a tent other than those stand-up garden shelters that protect you from sun and wind; or the child's den of a sheet over a clothes-horse. And here was the same adventurous make-believe sanctioned and indulged in by adults.

Two or more friends always came at Easter, so we had usually moved out by then, regardless of weather or inclination. And if not determined by Easter, then migration tended to start with the first signs of spring in the air. These are always false omens, but they lift the spirits of the most pessimistic of us.

By April that year there was still plenty of snow about, but trains were running between London and Carmarthen, and buses were getting through to Llangrannog. Their combined services were going to be bringing some very special friends to Penrallt for Easter: Stanley and Nellie and their daughter, Stella; and Felicité. All of these were camping people, but there were not enough tents on the farm to accommodate them so they mostly slept in.

I'm not sure if Felicité and I shared a tent that first holiday. Probably we did. I had heard a lot about Flicit, and it was apparent to me that she had for some years held the position of a sort of adopted daughter with Madge and Cecil – a position that I now seemed to be taking over. I was very concerned to avoid any appearance of conflict between us on this point and would retire into the background in any matter requiring precedence. Flicit was well aware of this. I remember her once saying, when she and I and Cecil were gong off somewhere in the car, 'Look, Cecil. Sheila has given up her place in the front seat for me.' For her part, she was confronted suddenly with a close relation of Cecil's, a young, recently bereaved orphan in need of love and support, who, she probably reckoned, should take precedence over her in the family hierarchy. For some time we behaved very delicately and tactfully towards each other, and so became good friends.

Felicité came often to the farm for holidays. She also spent two or three quite long periods of time there when she was between jobs. These might be for a few weeks or once it was nine months, which we all enjoyed. Eventually in the early fifties she gave up other work and moved in permanently, becoming a working and investing partner. She improved the accounting system considerably, but a lot more than hard work and money was required to make the farm properly viable in these more competitive times. My husband was a very clever farmer and manager and when he and I joined the business in the mid-sixties, he moved everything into the next level of mechanisation and easy management by installing (and mostly building) a milking parlour, and by starting silage-making and loose housing for the cattle. Also the number of cows doubled the day we moved in for we brought our own herd with us, making a total of 34.

In 1947, Easter fell in the first week of April. (I don't remember that. I looked it up.) Visitors were coming so we were having to move out among the snow dunes. Well, not quite; but the hedges were still solid with snow, and great bands of it, several inches deep, spread out into the fields. Somewhere around the first day or two of April, we set out to erect our annexes. And you might have thought we were off to sunny Spain from the jollity of it all.

There was always an excitement about the event, even for me. I suppose it was a milestone in the year, like the excitement of getting ready for the hay and corn harvests, which when you got into them were very hard work, worrying and frustrating.

I'm sure nobody else shared my disenchantment or knew of it.

On the first suitable sunny day in March or April, we would all trek off up the steps to the hayloft over the stable. As well as hay, all sorts of empty and full boxes were stored here, plus seasonal oddments like the rubber dinghy and incubators for eggs and young chicks. These mostly filled all spare floor space and had to be moved around and climbed over in order to get at the tents. Each was trussed to a beam like a sail to its boom. There was much cursing and sneezing and disruption of spiders as the heavy green canvas rolls were untied from the dusty, cobwebby beams. The long, cumbersome things then had to be manhandled through the small door, round a sharp bend and down crumbly concrete steps. They were put down on the yard, which David, or in later years William Henry, had swept clean. They were untied, opened out, brushed and shaken and checked for mouse damage and mould. Occasionally a mouse nest would fall out, beautifully made of twisted hay stalks with wisps of the soft hair lining twisted among them. Sometimes there would be a family of tiny babies tucked up inside. These disrupted families were quickly and warmly rehoused inside waiting cats.

The tents were set up in the Home Field, a smallish area of about three acres, behind the house. It was a field of passage between the yard and the north-eastern half of the farm, a place where the cows gathered to go in for milking, where the sows were turned out to exercise and rummage about, where the goose-house was. In short, a veritable mediaeval market place.

Positions would be found on the driest bits of land and away from the main traffic routes. The worst of the cow-muck was scraped away and tents were laid out, hoisted up, pegged down. Thick rubber groundsheets were laid inside, then blankets, pillows and sheet sleeping bags. There was something immediately cosy and intimate about crawling into that atmosphere of green-filtered sunshine with its subtle perfume of mice and horse and hay.

Inevitably it rained soon after; gales blew, temperatures dropped to near freezing, the land turned to mud and puddles. The groundsheets were thick and solid and I never knew water to come in from underneath, but it did come through the fabric of the tent itself because it was not entirely waterproof and leaked when it was touched. It was very hard to avoid touching it as you moved about. Later we had tents with flysheets. This extra layer, with a gap between it and the tent fabric, stopped rain from falling on the tent itself. Touching it was then quite safe.

Dutch barns supported by very rough tree poles. There is a clamp of roots in the foreground and some hay left in the shed behind. The cart stands in the other shed with its shafts on the ground, Welsh fashion.

I got used to the restrictive straitjacket of a sleeping-bag. I accepted that the flat, hard ground was probably very good for my back. I could even believe the hard clods and sharp stones were good for the soul. But the only time these conditions ever acquired the remotest semblance of comfort to me was when I had to drag myself out of them into the teeth of a force eight gale and pelting rain to adjust the ropes and pegs of the tent. These quite high, heavy structures were very vulnerable to rough weather, though if properly cared for, they stayed relatively upright and protective even in the worst gales. But to care for them, you did have to go outside, because what had been a well-constructed, well-pegged and tied-down tent in dry weather became a very different animal in rain and gales. Complete collapse was not unknown.

At times like these I would lie in bed listening to the canvas banging and the taut ropes squeaking and straining, and I would try to steel myself to get out and slacken the ropes before it was too late. The old-fashioned ropes shrank when they got wet and if not slackened could either snap or pull the tent over. The longer they were left in the rain, the stiffer the wet ropes became and the wooden toggle devices would wedge solid on them. Often the only thing you could do then was pull up the tent peg so you had some slack to juggle with. This gave you the extra and difficult job of banging the tent peg back in again. The soil was shallow and stony, making a firm pegging hard to find.

The tents were heavy and cumbersome to deal with in any conditions, but in the rain and wind, with nothing on but a cotton overall to cover my nakedness, possibly a coat of some sort if it had been really cold or wet at bedtime, it was not an enticing prospect. The job was ten times worse if one or more of the corners dropped down, so in the end it was best not to hang about. Pulling up the heavy canvas was like trying to pull a reluctant calf on a halter. Just such a dancing and prancing and dragging-back went on. A torch had to be balanced somewhere or gripped between the knees to show where the pegs went and how the guy-ropes should be fixed, and the toggles. And all the while, rain was not only soaking me, it was also seeping through the sagging material of the tent where it stuck to itself or fell against the support poles. Infiltration provoked by this pressure dripped through onto clothes and bedding.

It was not long before perfidious, weak doubts stirred under my robust enjoyment.

Part of the outdoor experience was to have tent flaps open all night with one's head all-but outside to get the full benefit of the fresh air. (My aunt and uncle were Girl Guide and Boy Scout leaders, often taking groups of children on camping holidays in the country. Poor little sods!)

One night I half woke to feel a slight drift of rain on my face and a waft of warm air. There was also a strong smell of cow, always stronger when it rained. Something slimy fell on my cheek. A slug! Urgh! I sat up, swiping frantically at it. My head banged against something hard. It was a huge, wet nose. A tongue like coarse sandpaper rasped forgivingly across my face. I flopped back onto my pillow and watched the great head swaying from side to side, masticating and dribbling regurgitated grass juices. Some dribbled onto my face. I decided I could do without the extra fresh air and I not only moved my pillow to the inner end of the tent, but from then on I also decadently closed the flaps against even light wind and rain. I was becoming more confident in my position in this new family, and I had these moments of rebellion.

Cows really were a hazard. Their noses were into everything – also their horns, hooves and teeth. They chewed at ropes and any loose flaps of tent. A horn could rip cloth and pull up pegs. Openings needed to be firmly shut by day otherwise belongings would be chewed and dragged around the field.

It was very soon deemed necessary to keep tents and cows apart and a tent enclosure was fenced off. A year or two after this, further decadence set in. Two custom-built chicken houses, complete with two wooden bunks and a chest of drawers in each, took the place of the tents. This was real luxury which even I quite enjoyed, despite the nightly trek through all weathers to get to bed. Into this enclosure too came Jimmy Burton's caravan. The Burtons, Jimmy and Sylvia, were friends of Madge and Cecil from when they lived in Kent and they came every summer with three daughters. They were jolly, lively people, and when they were on the farm there were always games of cricket being organised, especially on the beach. Everyone who wasn't working would pile into Jimmy's converted army truck, taking swimming things and food for the day and of course, bats and balls and stumps. It was not my idea of a day on the beach and I used to station myself as far out on the 'field' as possible in order to sneak off round the rocks or into the sea. All too often a shout would come from Jimmy to close in or 'catch that, Sheila!' before I could escape.

I enjoyed it all, that first Easter, even the camping. It was still clear sunny weather and it was the first time I went to Penbryn beach. I ran about, went swimming, played quoits, picnicked – completely naked for the first time ever on a public beach. For the first time ever anywhere! The cold swimming did not bother me. That was something I was used to. Neither did I have any conscious twinges of reticence about the nakedness. If that was how all these fascinating people lived, it was fine by me.

After the week or so of the Easter holiday, I dressed myself up once more in my smart coat and hat and my high heels and travelled back to London with Stanley and Nellie and Stella. I don't know what on earth they thought of my finery. I hope they weren't horribly embarrassed. We all trailed around the London shops for a while until it dawned on me they were wondering what to do with me. I asked what they would be doing now if I wasn't there and Stanley replied promptly that they'd go home. 'Well, why don't you do that,' I said. 'Will you be all right on your own?' Nellie asked anxiously. 'Yes of course I will,' I replied. I think Madge must have told them I was a bit of an innocent and hadn't travelled about much on my own. Which was true as far as it went, except that what little travelling I had done had been mostly on my own. I was very good at finding my way about, even if I couldn't remember the way back. I had endless self-confidence when I was on my own and was afraid of nothing and nobody.

Madge talking to Brocken by the pond.

I wandered round Woolworths for a while, and I stole a key-ring. I had got into the way of shoplifting bits and pieces when I was a child, but recently I had been trying to stop what had become an addiction. Now I said to myself: 'This is ridiculous. You don't even want that.' And I never did such a thing again. Nor, I think, did I ever steal anything again.

I went back to Ipswich to fetch the rest of my clothes and my bicycle from my kind friends, and I returned to the farm as an established member of the family.

Chapter 10

Everyday Jobs

THE first thing to be done on a winter's morning was to light the lamps. We all had torches to see our way up to bed and down again, and they gave light enough in the morning to see to get the lamps alight. Cecil always left these ready on the table the night before, together with the can of methylated spirits and a box of matches.

There was a knack, of course, and a routine to be followed. Like all paraffin-burning gadgets, the lamps were temperamental. Care and precision were required, and your full attention. The genie of the lamp always knew if you gave less than this and took his revenge. It was no good, for example, setting the process going then thinking you had time to nip away and start making tea. Like as not everything would then collapse or blow up round your ears.

First of all you had to get just the right amount of meths out of the little can and into the shallow gully inside the lantern. (Easy? You have just crawled out of bed – galloped if you're late – it's cold and pitch dark except for a small beam of torchlight.) The meths can had a long, thin pouring tube, bent into almost an S-bend, and the lamp glass would lift just high enough for the end to be eased through the gap. A quantity of meths was then tipped into the little channel round the metal tube to which the mantle was fixed. Meths is ideal for lighting vapour lamps because it burns cleanly and hotly, but if you poured too much it would run over and set other parts burning, whereas too little would not get the tube hot enough to do the job of vaporising the paraffin. Both situations produced flame that sooted-up the glass. Once the meths was alight, you watched and assessed and counted to about 20. Then you screwed down the valve and pumped up the pressure ever so slightly. If the mantle started to glow, you pumped a little more. If it didn't belch out a jet of flame, you pumped up hard until you had full pressure and a fine, bright light. Three lamps were prepared together, two to go to the cowshed and one to stay in the house.

David came in and collected the lamps at seven o' clock. He would be clearing the muck from behind the cows while Cecil had a cup of tea. Cecil then took a bucket of hot water to wash the cows and a cup of tea for David, and they would get on with milking. Tom must have been in on this somewhere too, but it's so long ago and I can't remember what became of him. I asked a neighbour and he couldn't even remember the family being there! Tom disappeared from our farm after a few months and I don't know if that was because his family moved away just then, or if it became unnecessary to employ him anymore because I was able to do a lot of the jobs he did. I was soon helping with the milking and Cecil and I did it together on Sundays when David had a day off.

Milking machines were installed before I arrived but I learnt how to hand milk because we had to take a few squirts of milk from each teat to test for disease before putting the milking cups on. Sometimes we hand milked afterwards to clear any remaining milk. So knowledge of the skill remained in the fingers.

The milking machine might seem an obvious labour-saving device, but like a great many farm machinery innovations, this was something of an illusion. It didn't make what you did easier, it simply allowed you to do the same amount of work more quickly, or do more work in the same amount of time. Which is not the same thing at all. Consider milking by hand. Once your hands learnt to squeeze and pull, it was not hard work – unless you had arthritis. The job could not be rushed. You sat on your stool and leaned against a warm, soft cow, and you pulled and squeezed until Cow said there was no more. You might need to run to empty your pail and wash your next cow, but then you sat down again. At the end you washed your pail, lifted the churn out of the trough of cold water where it had stood to cool the milk, and took it up the lane to the milk-stand.

It wasn't this ordinary and boring every day, of course. There were always sudden surprises like a tail swishing across your face or round your neck, warm and slimy with fresh muck. Or you might be walking along, well out of range as you thought, and suddenly you're caught in the machine-gun fire of a cow raising her tail and having a violent cough at the same time. Sometimes cows had sore teats, and if you were unaware and careless when you went to milk her, even the mildest lady would give an unpleasant kick and send you sprawling. If she really meant it, you could be knocked out. Teats are tender pieces of equipment and with even the smallest scratch they do not enjoy being squeezed and pulled. All quarters must be milked out if possible, both for the cow's comfort and to avoid the disease mastitis. The machine is gentler than hands and might be better tolerated, but there could still be a lot of kicking.

A cow's legs are assembled differently from a horse's and she stands up with her back legs first. A horse always gets up front legs first. This means Horse kicks her back legs out backwards, usually both feet together, while Cow kicks forward, straight into your milking position. We did have one cow, and only ever the one, who apparently thought herself a horse and would almost always push up with her front legs first. She would then be stuck in a sitting position. We had to get her to fold herself back down, then we pulled her head downwards while encouraging the back-end to rise up.

Milking-machine buckets were heavy and awkward. The lid, pipes and cups weighed as much as the bucket. But at least our buckets were aluminium and considerably lighter than the stainless steel ones I worked with after I married. Two, three or four cows were milked into one bucket, for they often produced only half a gallon or less each towards the end of a lactation, especially in the winter months. The milking cups were dipped in hot chlorinated water between cows to try to kill off any bugs. When the bucket was reckoned full enough to be worth emptying, the milking lid was transferred to the spare empty bucket and the plain lid from that was put on the bucket with the milk in. This was then carried out and emptied through the filter into the churn. Before taking it out though, you put the machine on another cow, so when you got back she was well on the way to being milked.

Soon after I arrived, a more efficient system of cooling was put in. The milk trickled slowly over two corrugated metal surfaces between which cold water continually ran – in at top right, out at bottom left. Milk cooled more quickly, thereby inhibiting the growth of bugs. Now though, you had to lift your milk bucket to head height to empty it. With one gallon of milk weighing 10lb (just under 5kg), you would be carrying and emptying weights of about 30lb (14kg) each time.

You will have noticed, I hope, the huge amount of washing-up accumulating in this new, improved system!

And while it all sounds vastly cleaner and more scientific than the old way, this was only true if the workpeople were clean and scientific. Think of all those cracks and

Bess, Flicit and Timmy bringing the empty milk churns back. The total amount sent was probably 24 to 30 gallons.

crevices, curves and corners, for bugs and microbes to lurk in. There were rubber teat-cup liners and their metal containers, various washers and small metal jiggy-bits, a collection of long rubber tubes and all those curves in the cooler. An army of scrubbing brushes was kept and constantly renewed. A swivelling cone-shaped brush fitted into the teat-cups; a hand-sized round brush scrubbed buckets; a brush with bristles cut to an angled edge scrubbed between the curves of the cooler; and various lengths of bottle brush went through all the rubber tubes and the little naked claws of the cluster. Everything had to be taken apart every day, thoroughly rinsed and put to soak in boiling, chlorinated water. It had to be scrubbed and rinsed, and later all put together again. After evening milking, things were just rinsed in cold water.

Being a milking machine, it needed something to run it, and this was done by a 1½hp

Lister petrol engine. These straightforward, solidly built engines were universally used for the job and were very reliable. Certainly I don't remember us ever having trouble with it.

The ultimate step in milking modernisation, as with all the latest farming mechanisation, has taken away practically all heavy physical work from the farm worker. Milk is now pumped straight from the cow into a cooling tank and straight from there into the milk tanker, which is driven down to the cowshed doorstep to collect it. Like all the modern systems too, it cost colossal amounts of money to set up. On top of the thousands of pounds worth of milking equipment, farmers also had to pay to have their lanes sturdily surfaced to carry the heavy milk lorries.

After milking, the machinery is set to pump hot water and chemicals through all the milking parts and then more clean water to rinse it all. This modernisation started to move into our area in around 1970 and Patrick, Flicit and I were milking with the system at Penrallt for the first time that year. However, in my first years it was buckets and churns and the milk being taken up to the milk stand by tractor or horse and cart.

During all the years I was involved with milking at Penrallt, there could not have been more than a few months when we were not on the early milk collection round. With both churns and tanker, that meant having the milk ready to go by 8am. This was very good for us. Made us get up in the mornings. Even so there were many times when we were later than was comfortable and I often ran out to put the horse and cart together with little, if anything, more than an overall on.

I loved working with Bess and soon learnt how to dress and drive her. In the winter she would be in her stable. She had a rope round her neck attached to a steel ring that slid up and down a vertical iron bar beside her. This allowed her a fair amount of movement, getting up or lying down, moving back and forward, tossing her head about. Thinking about it, she wouldn't have had much room to stretch out on her side, for there was only just room enough to walk between her and the wall, and that became very tight when she leaned against me. In the summer she had to be brought in here from the field, which could be a long and frustrating job. Even a bucket of oats might be snorted at if the grass was young and fresh and plentiful.

Having squeezed in beside her, I first put the great collar round her neck. This was turned upside-down to get the wide part over the wide part of the horse's head. It was then twisted around for that wide part to lie across her great chest-bones. That is where the pressure comes when a horse is straining to pull a load. The inside of the collar where it rests against her body is thickly padded with a sort of felt blanket material.

After the collar, the bit had to be enticed into her mouth and her ears arranged among the bridle straps, which were put over her head next. Then the heavy saddle was slung over and the girth strap tightened to hold it in place. From the saddle a wide leather strap stretched along the spine to her tail. Towards the end of this, four more straps hung down, one each side over the hips, the other two down the backs of the legs. These supported the heavy leather strap that curved round the hindquarters. Now we were ready to leave the stable and walk to the haggard for the cart. This involved going round several corners, which is quite awkward with a horse because she doesn't walk in a curve. She moves her feet sideways to point her body in a different direction and whichever way you are turning, you will be walking at that side of the horse's head, holding her bridle. As she feels you pulling her sideways, she will place the foot nearest to you precisely where your own foot would normally be. Most people get caught by this at least once, but a ton weight flat iron coming down on one's foot tends to stick in the memory as something not to be repeated.

I have not remembered all the technical details of the dressing operation. I have had to check with the fount of all such knowledge, my husband. Not only does he remember dressing a horse at the nearby farm he ran with his parents, he also remembers how it was done on a farm in Suffolk where he was evacuated at the age of 13! Apparently there was a difference in their way of harnessing the horse to the cart, and from the way Patrick described both methods, it was certainly difficult to understand why the Welsh farmers worked it the way they did. The obvious way seemed to be, as was done in Suffolk, to leave the cart with its shafts in the air with the chain that went over the saddle to support them permanently attached, one end to each shaft. The horse would be backed between the shafts, which were then pulled down to let the loop of chain drop into the groove in the saddle. The balance of the cart was now tipped forward, and this chain held the shafts in place. It just remained to walk round and do up the straps and hook together the bits that linked the cart to the harness, and horse and cart were ready to go.

The Welsh way was to back the horse between the shafts as they lay on the ground. The shafts then had to be lifted and held up while the length of chain was thrown over the saddle. Still balancing the shafts, the person doing the job had to shuffle round under the horse's chin and along to where the chain hung down waiting to be hooked to the shaft on that side. Perhaps when you were used to it, one way was as good as the other. Then the churns were loaded onto the cart and away we all went.

Taking the churns up to the milk-stand was Bess's main job in the years that she was

1970: the modern dairy with its 250-gallon cooling/storage tank.

with us, though the job was occasionally done by the tractor. It was in some ways easier with the tractor, some ways easier with the horse. If the tractor started when asked, this could be much quicker. The tractor would also stand still when told. Horses were not very good at standing still when told but were brilliant at starting. If the tractor started on the first turn or two of the handle (no push-button starters then), this was obviously much quicker and easier than dressing a horse and cart. It did go slightly faster up the lane as well. Both Flicit and I enjoyed taking a horse though, if possible, and visitors were always thrilled to ride up on the cart.

I helped Madge in the house quite a lot. Jobs that became my regular daily chores were the blackleading of the stove and the polishing of the floor tiles in the living room and hall. They were very old tiles, coloured alternately red and black. The glazing was a bit worn and pockmarked, but they had quite a good shine and this helped to give the room its warm, homey feel. I did other bits of cleaning and helped with this and that, inside and out, but those two jobs I somehow acquired as specifically mine.

I did most of the butter-making too, but that was no more than two, occasionally three, times a week. The butter churn was a square, glass jar that would hold one gallon; though only half this amount could be churned at a time. It had a screw-fitting metal lid through the centre of which ran a metal rod. On the top of the rod, outside, was fixed a gear and turning handle, and on the bottom, inside, was a small, three-bladed wooden paddle. Cream was skimmed daily off the bowls of milk and was collected and 'ripened' over a few days before being put into the churn and warmed slightly beside the fire. Or cooled in the summer. Perfect churning needed the perfect temperature. It was rare to get that. You sat with the jar on the table, or beside you on a bench, between your legs or on them. Nowhere was perfectly comfortable. If the condition of the cream was just right, including temperature, it would 'turn' in 20 to 40 minutes. Otherwise you could sit and turn the horrible handle for an hour or more. The job was fairly tedious and made the arms and back ache. On a good day the butter would suddenly be there. It might be so solid the paddle couldn't turn. If it was over-warm, it would set soft as whipped cream and if I was reading or listening to a play on the wireless, I might go on churning for some time without noticing it had turned. When the cream was too cold, a little hot water added at the right moment could set it. Overall though, I reckoned there was some other rogue element in the cream itself, or a gremlin in the churn, that determined whether you'd get butter or would go on turning the handle round and round forever like a pastoral Churning Dutchman.

All the regular outdoor jobs were to do with feeding and cleaning cows, horse and pigs. I did quite a lot of mucking out. It needed little more than a fair amount of strength, which I had, and an eye for building a well-balanced load. Cecil was always instructing me on the scientific and ergonomic ways of dusting and sweeping and he also explained how loading a greater amount on the front of a wheelbarrow allowed the wheel to carry the main weight, taking it off the handles and my arms. The animal houses were all uphill from the muck heap so you could achieve a good speed going down the yard. This provided a fair impetus to get up towards the top of the heap before much pushing was needed. The only problem was that once the muck heap had covered its base area, it had to be built upwards and it got quite high by the end of winter. The barrow had to be pushed to the top for emptying and for this a hard surface was needed. The traditional way to wheel wheel-barrows upwards, as you will see on any building site, is along a plank, and it was along and up one or more planks that ours had to be moved. It took a little time to learn this skill and several times I was forced to jump off into the bog to avoid being pushed or pulled deeper by an uncontrollable wheelbarrow. I got splashed, I got my wellies stuck or submerged, depending on how much rain there had been and whether I was wearing any, and I landed with my hands and arms half buried. But I never ever fell completely flat and only ever saw one person do this spectacular flop.

There were always calves to be fed with formula milk from a bucket using an imitation teat or sometimes just fingers. The babies were taken from their mothers within days of birth so that the cows' milk could be sold. This was done partly because formula milk was cheaper, but also because most of the calves were to be sold and must be accustomed to feeding from the bucket before they left the farm.

Then there were the chickens, geese and ducks, and the cats and dogs. And, of course, the humans. All had to be attended to one way or another.

Looking back on those days, everything was so leisurely compared to even 10 years later. Pressure became stronger and stronger on farmers to produce more and more. We were directed, advised and paid to grow specific crops using certain types and quantities of fertilisers for maximum yields, and cows were bred to produce greater and greater amounts of milk. The work became heavier and everyone had to rush about more to keep up with it. But in those first years there were always times when we could spend all day, between milkings, on the beach. We could walk and drive round the country; and we could sit and talk and read and listen to music.

Chapter 11
A Very Special Year

WHEN I came back to the farm that first April, 1947, bringing my very few portable belongings, Madge gave me a great hug and said, 'Welcome home, lass.' And so I was: home and welcome, and happy to be so. And so I remained for six and a half years, with two short excursions into other lives during the winters of 1947/48 and 1952/53. I married and left permanently in the autumn of 1953. Then, after a break of 14 years, I came home again with my own family to what was now my own farm. (Well, partly mine, in partnership with my husband and Felicité.)

All of that was to be donkey's years away though, and not a single one of those future years would contain such extravagant contrasts of life and weather, or such a variety of new experiences, as this amazing year of 1947.

The snow I came back to was not as white and shiny as it had been when I'd left two weeks earlier. In spite of brilliant sunny days – or perhaps because of them since they brought sharp frosty nights – the heaps shrank very slowly. Those by the roadsides became splashed with mud and blackened from exhaust fumes. The fields were much the same. Only here and there on the high banks and in remote corners were there still unblemished stretches, smooth and sparkling.

'There's a surprise for you,' Madge said as we stepped down into the sitting room. I stood a minute, looking round the room. I saw two people I didn't know sitting on the settle in the chimney corner.

'This is Hilda.' I went over and said hello. 'And her mother.'

I had heard a lot about Hilda. She was a friend of Madge's from wartime days in Bridgend. She had been to the farm a few times and had helped with house decorating and scrubbing and some stooking of corn in the fields. She was a lively lady, kind and courageous and fun. Through the 30-odd years that I knew her, she was like Fate's Aunt Sally. Time after time she was hit by disaster, tragedy, treachery, and drudgery; and always she bounced back up with a smile and a joke. Though a couple of times that took all her guts as well as a push and a pull from her friends.

'Hilda knew your mother,' Madge said.

I stared blankly, then smiled a sort of second greeting to acknowledge this closer acquaintance. I was trying to think how they had known each other. I went through names of friends from a long way back that had come up in family conversations sometimes. There had been a Hilda, but I had met that Hilda only last year.

'We knew Hilda came from south-east London,' Madge said, 'but talking about old times last night we found she actually lived in Lewisham, which is the next borough to where Cecil lived. Cecil said he thought your mother came from Lewisham and when he said her name, Hilda's mother said, "I knew the Lannigans. They lived two doors down the road from us".'

'We all played and went to school together,' said Hilda.

I stared at them, thoughts about all these people in my past and present tumbling about in my head like clothes in a washing machine. It was a very strange coincidence. We were people from three entirely detached families, who had come separately to Wales, met up bit by bit, then found we had these strangely interwoven links between us.

'She looks like a Lannigan!' pronounced Mrs Arnold. Which made us all laugh, for some reason. Hilda remembered my mother, Alice Lannigan, but she had known my Auntie Kathleen better because they were the same age. Hilda and Kathleen met up some years later and had terrific laughs and remembrances about the old days. They made interesting comparisons of their later lives and all that had happened to them in the years between. They were the same breed – tough East End ladies who had lived through two world wars. Like Hilda, my Auntie Kate sustained many losses and disappointments in her life but was always cheerful in between. She was a brilliant manager of a very small budget from which she somehow saved money for holiday treats for her daughter and for my brother and me. She always found time and infinite tolerance for those she loved. When she occasionally felt put-upon, she would prowl around in the style of the old melodrama villains, singing in a deep, sinister voice, 'If you want any Dirty Work done, send for me.' She knew all the old music hall songs and dances and did a splendid Marie Lloyd impersonation. Cecil also performed some of these old songs. They would have made a great double act, but sadly never came together in this way.

Hilda had a son, Ken, who was about the farm from time to time in early adolescence and he appears in this book in stories and pictures.

This was the first, and probably the most extreme, of three very strange coincidences that occurred in those first years in Wales. The second was a year later at a village concert. I was standing at the back of the church hall with a few other young people when my eyes focused on the back of a head about halfway down the hall. I stared, possibly gawped, in shocked disbelief at the clean sharp parting in the back hair where it was swept smoothly aside into two neatly coiled, earphone plaits. There were no stray hairs or untidy protrusions. All was as neatly disciplined as it, and I and my fellow pupils, had always been, at the Ipswich High School for Girls under the direction of Miss Neale, our headmistress. Even from the back, and even though quite a lot of older women still wore their hair in this fashion, there really was no mistaking this particular head. It had to be Miss Neale. But still I waited to see her face before flopping down with a gasp on the wall by the door. I must have made quite a fuss for one or two boys in the group asked if I was all right. I really was quite shocked. This place was so remote from anywhere, but especially from so far away as Ipswich. When I said incredulously, 'I've just seen my old headmistress in there!' one of the boys asked if I was afraid of her. I told him no, I was just astonished to see her here. I think they thought I was daft. They couldn't see why she shouldn't be here. In fact it was not so very astonishing. For one thing, Llangrannog had been a well-known and much frequented holiday resort in certain circles of society for many years before the war. For a second, the little town of Llandysul, about 15 miles away, was the home town of the school secretary. But neither of these things did I know at the time.

The third coincidence didn't come to light until 1953 when I met my husband. His family had bought a farm roughly three miles away from Llangrannog at the same time as my aunt and uncle bought Penrallt. He and I met in 1953 and discovered that we had been born roughly three miles away from each other in Essex. What complicated puzzles and patterns the Fates work out and weave into their tapestries of our lives.

Soon after I came back that April, I had my 18th birthday. This was not the special event it is these days. Twenty-one was the big one then. It was my first experience, though, of Madge's special way of celebrating a birthday. Apart from the usual things like cards, presents, cake, drinks, I found when I came to the table for breakfast, a decoration of small, fresh flowers arranged in a ring round my place setting. Madge did this for everyone who had a birthday at Penrallt. It was a very pretty custom; very special. She even found enough bits and pieces for a garland for Felicité when she was with us for her birthday in mid-November. After that introduction, I always helped to prepare the garlands and I also made them for Madge on her birthdays.

Another nice thing Madge did was to make little plate gardens with plants and odd leaves and flowers from the hedgerows. These arrangements are commonly available in flower shops now, but not then. We prepared a bowl of stones and moss and stuck in twisty twigs, bright berries, cuttings of pine needles, sometimes tiny trees. When the snowdrops appeared, a clump of them would be transplanted for the main decoration but it was the adding of other shapes and colours that turned a bowl of flowers into a miniature landscape. I loved collecting the pieces and putting them together and I enjoyed looking at the finished display. I made up stories about creatures living in these

Winter view of the cliffs to Lochtyn where there was an Iron Age fort (500 BC) on top of the mound, Pen Dinas Lochtyn. We often walked across to the Ynys (island) and peninsula, for the beaches and gull's nests. As far again beyond is Cefn Cwrt. Emmy's Rock, half submerged, is in the centre foreground.

small worlds. I suppose it was a pastime equivalent to computer doodling today, though I find myself thinking I sound more like 10 than 18 years old!

All of us were much younger in our ways and knowledge in those years, but even compared with my peers I was astonishingly ignorant and naive. Among a collection of old notes and letters, I have found a postcard sent to me when I was just 17. It's a traditional pastoral scene and in the foreground is a small cherub child, part covered in a sheepskin and holding a shepherd's crook in one hand and cuddling a sheep with the other. This postcard was sent to me in the office of the factory where I worked, by young men from the Drawing Office, away on holiday. On the back it said, 'A picture of innocence for the innocent.' I was not the least bit put out. My friends were more outraged and embarrassed for me. I showed it to them quite happily with the comment that it did look like me, didn't it? Only now I cringe and smile ruefully at the memory.

As well as making gardens for ourselves, Madge also sent them off to friends in town to brighten their concrete days. I sent them to my friends and relations too.

Thinking of packing them reminds me of a strange, old-fashioned custom we had with parcels. Madge and Cecil had sealing wax which we melted onto the knots. It was not something I had encountered in my lowlier time of life, only in books and historical films, so I found it fun to melt this stuff and actually imprint a seal onto it. The practice must have been almost extinct then, but we kept doing it until the sticks of wax were used up.

I don't remember much about the farming jobs that year. I probably did small bits and pieces of things like raking the hay together, leading the horse, helping with fetching and carrying. Stooking, of course, come corn harvest, and forking the sheaves on and off the carts and trailers that took them out of the fields and home for stacking.

Mostly I helped Madge with cleaning the house and preparing and serving food. Visitors started to come for holidays, most of them friends of Madge's. Madge had what seemed to me countless numbers of friends. I still find it difficult to imagine collecting and keeping so many, finding time, memory and bigness of heart and mind to care for them all – us all. I met all the oldest and closest of these friends that summer, plus some who were not yet so well known but became very close later.

Most afternoons when we were not harvesting and there was nothing else particular going on, I walked down to the village and explored the beaches and the cliffs. I swam and climbed and sat dreaming and composing sentimental poetry and stories. There was one of those occasional very low tides that year and for a few days it was possible to walk along the sands at the bottom of the cliffs all the way to Ynys Lochtyn. This is about half a mile I would think, maybe a bit more. There were endless shoals of mackerel all along the coast that summer. At one time people were scooping them out of the sea within paddling distance of the shore. A few even flopped on the beach and were stranded there. I walked out to the Ynys one day and found one of the village boys fishing off a rock using snippets of silver paper on his hook. He showed me the dark patches in the sea that were great masses of the fish. He was pulling them out as fast as he could cast his line and unhook them. He offered me a go and I did catch some, but it seemed to me a barbarous thing to inflict that hook on a little animal. I could have swum among them with a net catching all I needed much more humanely, and probably more quickly, if I'd thought of it.

Another wild food I learnt to gather was seagulls' eggs. Madge and Cecil had recently spent some weeks on Skokholm Island off the coast of Pembrokeshire, working with the naturalist Ronald Lockley. There they learnt that if you take eggs from a gull's nest it will go on laying, within a set period of time, to replace them. Therefore it was no great hardship to those birds or the breed in general if people collected and ate some of the eggs. And this is what they did, and what I learnt to do as part of our living. We used a dozen or two or three a week during six weeks or so from the end of March to early May. They would stop laying then so if all their eggs were taken they had no young. They don't go on for ever like chickens. Their season runs out after a few weeks because then it becomes too late for chicks to hatch and learn to fly and feed themselves before winter sets in.

It was necessary to watch the nests from late March onwards in order to collect the very first eggs of the clutches. This ensured the eggs were not being sat on and the embryos starting to form. I collected from a wide spread of nests and was always careful to stop collecting in good time for the birds to lay and hatch a full clutch of eggs.

To start off with we all walked over the cliffs with our rucksacks to Ynys Lochtyn, where hundreds of gulls nested in more or less inaccessible spots. I learnt a lot of my climbing skills while venturing farther and farther down and along the cliffs and ledges in search of gulls' eggs. Mostly I did this on my own as Madge and Cecil became more tired and tied by the farm. It was something I loved doing: the excitement and skill needed on the treacherous slate cliffs, combined with the views, the birds, the useful product to take home at the end, all are very clear memories.

People expect gulls' eggs to taste fishy but they don't. Mind you, I never tried one plain boiled. They were very big and soft-shelled; and I suppose there always was the possibility I might have picked an egg from a nest that had been sat on and so there could be an embryo seagull inside. Madge whisked them into mounds of fluffy scrambled eggs and splendid omelettes. And quite often we ate them with another new 'Food for Free' which I went gathering. Stinging nettles. I thought Cecil was playing one of those idiot jokes on me – 'go and ask for a tin of striped paint' – when he said 'What about getting some nettles for lunch?' But I did as I was bid and then obediently ate them. They were as nice as any other green vegetable, all of which I like anyway. They have one big drawback. As with spinach, a large saucepan full cooks down to a couple of tablespoons and it takes an age to pick the bucketful you need to feed four people. For unlike spinach, nettles are small and fiddly to pick, and they sting! You nip out the top four or six leaves only, though others can be picked off the stalk if they're fairly young.

The brilliant sunny weather went on and on. After the exceptional winter, we had an extraordinary summer. My husband says there must have been some rain because the crops of hay and corn were so good, and I do remember now some wet windy weather in June. The trigger for this comes from the memory of sharing a tent with a young pregnant cousin. The blue skies and sunshine that had blazed over our snow banks shone on through most of the haymaking and corn harvest, picnics and beach parties, and the carnival.

Madge was working her way into the politics of Llangrannog and she decided we should join in the fun of the carnival. I am fairly sure no one from the farm actually participated that year. Except me. I went as Summer 1947, dressing myself up in every kind of wild flower I could weave into the costume. The main item of this was a ragwort skirt with maybe two dozen plants in it. They were all tied together with binder twine. It was very heavy. Trying to remember the feel of it I would say it must have weighed at least 10lb (5kg).

Ragwort is a magnificent looking plant with huge yellow heads of flowers. It is also a strong, noxious weed that rampaged through fields and hedgerows, seeding prolifically. It was a great menace, crowding out crops and being poisonous to plants and beasts. Today it has been almost eradicated and there are penalties for farmers who fail to clear plants from their land. Councils too must remove any plants that appear on roadsides and other public places. When I was a child in Suffolk, I remember seeing quite puny ragwort plants around the wilder bits of recreation grounds. I noticed them because they would be completely covered in yellow and black striped caterpillars, all busily eating any fragments of leaves that were still left on the plants. These creatures are also, sadly, almost eradicated. I have lately found a few on plants in private woodland, and in forestry land, and have made a point of not pulling them up. Though

I fear that other conditions besides the destruction of their food plants are driving the butterflies away.

I joined the parade in all my glory but didn't win anything. They were a wild and motley lot, dancing along, shouting and singing and playing various instruments like toy whistles and saucepans banged with spoons. I remember only one character clearly. She was a largish, laughing lady in a voluminous black dress which was tied round the middle with twine. She might have had a cushion under it. She wore a hat of some sort squashed down on her head and she was pushing along an old pram with a gramophone in it. She stopped from time to time and wound the gramophone up. I don't know what it played but when she stopped, the whole troupe stopped. It took long hot ages to drag on down the road to the judging place outside the Ship Inn.

That was my introduction to Mrs Owen, a kindly, slightly outrageous lady, who I became very fond of over the years. One other particular exploit I remember about her was when she wanted to go up the road to the shop and couldn't find her hat. She seemed to have an obsession with hats and wouldn't go anywhere without one. She turned up in Mrs Thomas's shop on this occasion wearing a rather unusual beret-like creation. Mr Thomas, who was a straightforward man with a dry sense of humour, made some comment about Paris fashions coming to Llangrannog. Mrs Owen laughed, took it off and gave it a shake. It was her son's bathing trunks, pinned into shape with a large brooch.

Sheila, Madge and Hilda having fun in the dinghy.

Chapter 12

Not Suburbia

MADGE was oppressed from time to time by the amount of mess everywhere and her inability to keep the house looking as nice as she would like. We cleaned and polished and scrubbed, but there was always dust blowing in when it was dry and mud and muck tramping in when it was wet. Cecil tried to persuade her it didn't matter.

'It's not suburbia where people have nothing better to do than poke their noses into other people's houses and count how many grains of dust they can find.'

Mostly Madge accepted this but it was annoying when we'd spent time and effort getting everything really nice and then some emergency brought muddy boots tramping in. Normally all boots were left outside in the porch, as were the overalls we all wore for outside work. These were not onerous habits to get into and made a tremendous contribution to keeping the house clean, allowing Madge to get out and do other things. It was not that she was fanatically house-proud. She was just very good at making a bright, comfortable home, which we all appreciated. Nobody minded putting wellies and overalls on and off to keep it that way. The living room furnishings were bright, basic and easy to clean. Over the years the chairs acquired loose covers, so they were lighter and cleaner.

Another conciliatory gesture from Cecil was to get David, and later William Henry, to sweep the yard clean every Saturday morning, unless other jobs of extreme urgency needed doing. I don't know how they felt about this, but the job was always done thoroughly and diligently and it did make a great difference to the general cleanliness round the house.

Everything and everyone went in and out through the living room all day, going between the outside and the kitchen with vegetables, milk, cans of water. Then there was fuel to come in for the fire, hot water to go out of the boiler and cold to be tipped in. It was a well-used highway.

The sitting room, though, was kept for special. Not so special as it had been when it was a Welsh Parlour, but used entirely for relaxation and not to be gone into in working clothes. It had a lovely bright green carpet and a red three-piece suite. There was a useful, two-foot square polished wood table, a matching bureau and a cabinet four or five feet high by three feet wide, full of gramophone records. We couldn't use the magnificent radiogram because there was no electricity, so that stood in a corner unused until 1951. We did have a wind-up gramophone and that was greatly used. We spent many winter evenings and wet afternoons listening to music. It is surprising to think how little irritated we were by having to wind the gramophone and change the record half a dozen times during a symphony or concerto. Some of these sets of records had automatic coupling, allowing you to stack three or four or more together on a special spindle. They played one after another giving us the illusion of listening to a real orchestral performance for a whole half symphony before having to get up and turn the records. The magic had to be stretched a bit at the end of each record's playing for then there came a little whirring noise, a pause, a slight scraping and then the most tremendous crash as the next one dropped down. This luxury had to wait for the coming of electricity though. The ordinary gramophone couldn't do that trick.

It must have been a very complicated and frustrating job for the recording musicians, having to arrange these long pieces of music to fit onto a precise and limited number of records. The breaks could be quite abrupt, speeds of playing sometimes very strange and erratic. To save money, a piece might gallop along at tremendous speed in order to make it fit onto the minimum number of records.

The gramophone had the additional irritating distraction of having to be rewound several times during play. Going back to such a performance now would drive me mad, as do the atmospherics and other poor reception noises we occasionally still get on the radio. Yet we used to listen to and enjoy music through the most awful cracklings and whistlings those years ago.

The front door was always left open during fine weather and one day the inevitable happened – a calf got loose and walked into the house. No one noticed for a while, but a sudden shout from Madge brought me running. Madge had gone into the living room from the kitchen and found the calf nibbling the tablecloth. She gave it a whack to send it out and it was on its way through the hall doorway just as I came barging in. It jumped back and skidded on the tiles round the armchair towards the kitchen. With remarkable agility, Madge spun round and got there first. I was also running round the chair to head it off and this left the way clear for it to trot back and out through the door to the hall. Only instead of turning right to go outside, she went straight across into the sitting room.

'No,' Madge wailed, 'She'll ruin the carpet. She'll break something.' She turned on me. 'Why didn't you shut the door? Bloody stupid coming in here to get a calf out and leaving the sitting room open.'

'I didn't know she was here. How should I know you'd got a calf in the house?'

'For Christ's sake don't frighten her now. And if you see her tail start to go up – ' she looked around and grabbed the vase off the hall chest, hastily throwing the flowers and water through the front door, 'stick this under it.'

The animal was standing in the middle of the room and seemed to be looking out of the window. She was not a baby calf but about three months old. Too big and strong to womanhandle easily. As we crept closer, we saw she had one of the embroidered chairback covers dangling from her mouth, which she chewed and dribbled over. We walked carefully round her, quietly telling her the most awful lies about what a lovely good girl she was, heaving against her flanks to push her towards the door. I took hold of the chair cover and tried to ease it out of her masticating mouth. We had her almost at the door when Cecil turned up. He backed off immediately, but not before the startled calf had leapt back onto my foot and barged into the table.

'Get a bowl of corn, Cecil!'

'Hurry! She's going to crap.' I felt her tail moving under my hand.

'Get her out!' Madge roared. We both thumped her on the back, banging and shouting and pushing as hard as we could – a policy guaranteed to scare the innards out of any self-respecting bovine. She jumped forward and her tail went up. The important bit was still pointing straight towards the middle of the sitting room. I thrust the vase forward, hole to hole, tripped over the new smart doorstep and sprawled in the hall behind her. All I could think was, 'thank goodness she's out of the sitting room'. And then: 'and she hasn't crapped on me!' I became aware of Madge laughing. I sat up and saw her collapsed on a chair, howling with laughter. There was no muck anywhere. I looked through the front door and saw Cecil out on the yard. He was standing with one arm round the neck of the calf, his other hand holding a bowl under her nose as he soothingly told her what a poor little Funny she was and did the nasty aunties upset her then. I began to giggle.

This calf was named Sheila. I'd had to wait a long time to have a calf named after me for others were in the queue before me. When she did come along she was nothing but trouble. No other cow bothered about the washing blowing on the line. Only Sheila was found with knickers on her horns, a sheet wrapped round her middle, socks chewed and trampled in the muck. The clothesline had to be moved to a much less convenient place, all because of Sheila. Only Sheila broke through every hedge and fence, despite all kinds of barbaric devices designed to inhibit this. She had a heavy length of wood suspended from a rope round her neck at just the height to hit her knees if she tried to climb a bank. For a while her head was tied to a front leg, not tight down but just so she couldn't lift it high and pull out bushes and turfs. Another deterrent was to hobble her, one front leg to one back leg. These were all suggested cures from local farmers. None of them worked. I think she must have been reprieved many times beyond reasonable patience only for my sake. The final blow was when she refused to become pregnant. After two failed inseminations, Cecil had the vet check her and she was found to have some defect preventing reproduction. A very small girl staying with us at the time went to the cowshed to watch the examination. Afterwards she came in the house and announced, 'The vet's here and Sheila's very upset she can't have a baby.' She must have been a bit hurt by the roars of laughter this caused, not knowing I (supposedly) had a crush on the vet.

The doorstep I tripped over had been carved and set in place very recently by a German prisoner of war. These men were still in camps round the country as late as 1949, waiting to

Sheila-calf at a restrainable age. Visitors of unknown origin: Victor, Mabs, Donald.

be sent home. Many of them didn't want to go and found homes by marriage or by getting permanent jobs. They earned a little extra money in the meantime by doing casual work, mostly on farms. They were brought out from the camps in army lorries and dropped off in twos and threes on farms where extra hands were needed. They were intended to help out with fieldwork like planting and lifting potatoes, or hoeing and muck-spreading. Many of them were craftsmen and these other skills were often more welcome than the field work. One of the Germans who came to us was a carpenter and he replaced the worn slate in the doorway of the sitting room with a beautiful hardwood step. It not only looked better, but even more importantly it removed a big draught hole.

The day came, I don't remember when but fairly early on, when Madge decided it was time to do something about the back kitchen.

'We'll get the whitewash off the walls, then cement them.'

So we did that, Madge and I, our tools mainly bread boards and carving knives. The knives were good for digging layers of whitewash off, not so good for applying cement. The layers came off very satisfyingly in great sheets to start off with. They were crisp dry sheets, stuck close together but still with curves and air gaps amongst them, a bit like cooked filo pastry. The overall thickness of these layers of whitewash was two to three inches (6 or 7cm), and this was on every wall in the kitchen. We filled sacks and buckets with the stuff, then wheeled it out in a barrow to the muckheap. I'm fairly sure

A harvest tea outside the house, with helpers from the village: Bidgi, Dai and Gwyn on the seat against the wall.

we took it through the house, which means Cecil had not yet put in the back door to the garden. The weather was fine, I remember that. It would have been awful in the wet.

The surface of the whitewashed wall had been fairly smooth and flat. All those layers levelled it over the years. When we took them off, we were back to the lumpy boulders. Madge had ideas of building up a flat cement surface at first, but it soon became obvious we weren't going manage that. It took far too much time and cement. I remember it was more than usually difficult to make the stuff stick because lime remained on the rocks and it sucked water from the cement before it could set. We mixed and carried, slapped and smoothed. Cecil provided a very small cement trowel, but, after all, we found the long, wide kitchen knives suited us better.

We were quite pleased with the result, especially after a shiny cream paint was applied over the lumps and bumps. We gained an extra five or six inches width to the room, which was a significant amount in so narrow a workspace. I was never aware of what the effect of this room on civilised people might have been, but many years later when my daughter was getting married, I took a photograph of her cake standing on a table in this kitchen. The place looks for all the world like the inside of a cave.

The outside walls of the house had been dressed with an oil-based substance of some sort and it took many years for this to clear entirely. The oil interfered with the sticking power of early colour washes which soon flaked off, but still the effect was dramatically more cheerful, especially with brighter paints on the woodwork.

Madge was now getting tubs of flowers outside the house. The first she had were old oil drums cut in half and painted commercial green. They were filled with bulbs in winter and then bright coloured summer flowers. It all helped to break up the rather stark austerity of the yard, making a welcoming glow. Gradually window boxes appeared as well and hanging baskets. And just to make sure Madge didn't sink too deep into illusions of suburbia, every now and again a cow would amble across the yard and clear a whole tub in one mouthful. Sadly she never took any photographs of these feasts. It would have made a lovely picture.

In 1948, three gentlemen friends of Madge's came to stay and they were organised into setting-in a window over the sink. This proved a great joy, giving us light, a view and a ledge to hold soaps and the like. Madge managed to get hold of some glazed tiles. I expect such things were scarce and maybe these were left-overs in some shop corner. Some were white and some were green, but the men didn't notice this until they had stuck down the green majority. They then had to put two or three odd ones on the ends of two front rows instead of making a pattern with them.

Of these three men, I got to know Arthur best. He was fairly elderly – over 50 anyway! – and came regularly for holidays. What is more, he came with a lady who was not his wife, though he did have a wife. It was one of those situations I had never met up with before and in so far as I had heard of it at all, I classed it, along with Love, as one of those things that happened only in films and other fairy stories. Of course, there were Madge and Cecil – which goes to prove my point, for what were they if not a fairy story?

Arthur had a favourite saying: 'Glory be to God!', which appeared frequently in his speech. I wrote this poem in praise of him and the window.

> The window's set in place
> For which we say the grace,
> Glory be to God!
> It's slightly on the skew
> But there's a lovely view
> Of garden, tree and sod.
> Glory be to God!
>
> The window ledge so wide
> Is filled from side to side
> With every odd and sod.
> And so it's rarely seen
> The tiles in white and green
> Are really slightly odd.
> Glory be to God!

I can't remember the last verse, only it ended by saying that glory in this instant was not due to God: – 'But to a mortal rather. Glory be to Arthur!'

Chapter 13

If Mrs Jones Can Do It, I Can Do It

BARNYARD fowls have always been, traditionally, the work and perks of the farmer's wife. Every worthwhile children's book used to have a farmyard story with pictures of a chunky lady in overall and wellies, a metal pail on her arm and a flock of assorted birds round her feet. This was Madge.

Feeding them made a pleasant walk out on a fine day and ensured she left the house for a while. It was one job she did not readily delegate, though visitors might be allowed to help sometimes. It was an easy, satisfying little job for townies and children, and there was the additional special delight of collecting eggs.

Charles particularly enjoyed calling the birds and leading them off to feed in the field. He even worked out one of his silly tricks to play on them.

First of all he rattled the food in the pails and called the hens into a group by the main barn door. While they clucked expectantly and pecked each other excitedly, Charles would walk through to the other side of the barn and creep out the back door. When he appeared at the top of the yard, the birds immediately recognised him and there would follow a mad race between all of them and Charles to be first to reach the feeding troughs in the field.

But they soon learnt to anticipate the trick. Some would scurry up the yard the moment the barn door closed. Some would go straight round to the back door and wait. Charles just strolled up the yard then, like the sober city gent he was. When he and Win went home after their holiday, the birds soon forgot the game, which started up all over again at their next visit.

Four o' clock was feeding time and like all workers the chickens had built-in clocks set for teatime. If feeding staff were late arriving, a deputation would turn up on the house doorstep, creating a disturbance and leaving unsavoury calling cards. In winter they gathered early, before the grey evening light turned to blackness.

Free-range chickens, in this old-fashioned sense, were a great nuisance. They had the run of the farmyard and buildings and they dropped their muck everywhere. On the yard it was just part of the environment but in the buildings there were tools ideally placed for perching on or perching over. You went to grab a hammer, or a saw or broom, and found your hand slimy and smelly with bird mess. (I know everyone says shit these days, mostly without recognisable meaning, but it's a word that was almost never heard when I was young and certainly not in average mixed company. So I stick to the euphemisms of the day which at least provide a meaningful description.) Another associated problem was that the tools rusted. The birds got in the hay too, which was fouled, likewise grain and other feed and seed in the barns. Doors were supposed to be kept closed, but this was obviously impossible when work was going on there, and they were inevitably forgotten sometimes as well.

And so enclosures were made for the chickens, which incidentally made egg collecting easier and more reliable. In a lot of ways it made life easier and more reliable for the hens too. They had comfortable, handy nest boxes and immediate shelter from the weather. Enjoyment of the Great Outdoors among hens is similar to that of humans: it's alright while the sun shines but come rain and high winds, most of us rush for cover if we can. Chickens hate rain and wind and they run as well. Enclosures also protected them from passing foxes.

Another convenience was being able to check them more readily for diseases. Contrary to most supposition, these farmyard fowls were not very healthy. There was inherent TB among them that would be spread about the place in droppings and the occasional dead body that could lie hidden forever. There were other diseases as well which mostly went unchecked. So while utterly against battery cages, I consider that other restrictive systems have advantages for both sides. Sadly, in all situations where people (and other animals) have power over each other, cruelty and neglect are likely to occur, however rigorous the laws and inspections. It is in any case impossible to farm chickens of any kind in the quantities required for consumption and profit these days without enclosing them. Only small self-sufficiency farmers can still share the pleasures and pains of living in close communion among their beasts.

Madge reared small batches of cockerels which were killed for home eating and selling during the summer, but the main crop was for Christmas. They grew first of all in an enclosure, then for a few weeks through September or October, depending on when the corn crop was cleared, two chicken houses would be moved into the stubble fields where the birds spent happy days scratching for corn, weeds and insects. The drawback to this was that someone had to walk across every evening to make sure they were all in their houses and then shut them in to keep them safe from foxes. For their last six weeks they were back on the yard where they were fairly closely confined and fattened. I think we prepared around 50 cockerels, plus up to six ducks and maybe four or five geese.

Plucking took place about eight days before Christmas. Annie Lloyd, as always for occasions like this, was in charge. She organised extra hands to come from the village and there would be six or eight of us plucking. We sat in the out-kitchen which was cold and dark and full of smoke. A fire was lit in the corner under a big open chimney hole but we could never decide whether it was better to freeze to death while it died right down and stopped smoking, or be slightly warmer and choke to death. We must have looked like figures from a Dickens low-life hovel in this gloomy room with our heads wrapped up in scarves, overalls tied round the neck and middle with twine, sacks over our legs for warmth and to help keep the feathers and fluff away.

David or Henry did the killing, and Annie, with help now and again, did all the de-gutting and dressing. Madge helped with the final tidying, singeing, wiping their insides and packing the giblets in, then sewing legs and wings in place. She then weighed each one and laid them out on the dairy slabs to cool. A note of their weights was put beside each. Next day we sorted and packed them according to sizes ordered by people or sizes of families who were having them as presents. They were beautiful birds, plump and tasty, ranging in weight from four to over eight pounds (2 to 4kg), though the majority were between five and six pounds. We packed and posted a dozen or so parcels, some containing two birds, to friends and relations all over the country. They mostly arrived in two days at most and kept perfectly well for Christmas.

Ducks and geese still roamed free about the place. These did not fly much, nor roost, so they were not the same problem as the hens, though they did still leave a lot of mess about. A big problem with the geese was what they did to corn ricks. It all started off harmlessly enough with pulling bits of grass and corn out. Then gradually a hole grew in the stack big enough for a long beak to reach in and find the juicy fat grains in the head of the sheaf. A pull and twist of those strong necks soon had the whole sheaf out, leaving the stack vulnerable to undermining and collapse. Chicken wire had to be bound round the stacks and along the sides of the ricks to a height of three feet to protect them when geese were around.

They were not the only things needing protection, for geese are very aggressive, frightening creatures. This applies to the gander mainly. The lady geese would hiss and swing their necks when they had young, but they wouldn't advance very far. The gander would charge anyone anytime. As he ran forward with hefty neck stretched out, open beak hissing ferociously, and wings at full stretch and beating, he really was frightening. We kept well clear of him when he had a family to protect. At other times the rule was that you looked him steadily in the eye and showed you were not afraid of him. And you never, ever, turned your back – not within 20 yards of him anyway.

Mostly the strategy worked, but one day it didn't.

I was going down the yard as he was coming up. I moved over to give him as wide a berth as possible – no point asking for trouble – and I kept my eyes fixed firmly and fearlessly on his. He was obviously having a bad feather day. I could see it in every quill and pinion. He had no intention of playing by the rules and was not going to drop his eyes. Two can play at that game, he said. No good running. He could move with amazing speed and it was better to face a running frontal attack than an accelerating take-off at my back. In any case, he made up his mind faster than I did mine and was already on the move.

With pounding feet and stretched wings, neck and beak, he came straight on target like a torpedo. He homed in on my thigh and took a great beakful. Not only did he get through the half-inch thick WAAF trousers, he got half an inch of flesh as well. He drew blood and left me with a bruise from knee to trunk which remained colourful for weeks.

And what was I to do next? I made a wild grab and caught him round the neck. My thumb was wedged under his beak and I held him at arm's length with his toes just tickling the ground like a dancing puppet. We glared at each other and I knew perfectly well that he'd won, even if he didn't!

So what now? I didn't know.

I looked around for help and saw William Henry and Ken standing a few yards up the yard, laughing their silly heads off.

'What shall I do?' I squawked.

'Swing him round your head a few times and let go,' said Henry.

'I can't do that!'

Nobody offered any other help.

In the end I edged towards the front door, still holding him by the neck. At what I reckoned a safe point I turned him away from me, gave an extra squeeze to his neck and a quick shove with my wellie, and ran for the house. I suspect I did not take my boots off before getting inside and closing the door. Though you never know, habit and authority override all sorts of other sensibilities.

I never had any other trouble with Mr Goose. Perhaps we were both more circumspect in our future meetings.

Mrs Goose produced, each spring, about 20 eggs in all. We took six or eight of these from her first layings, using them for cooking or we might give some to other people to hatch. Mrs Goose would then continue laying until she had a sitting of 10 or 12.

Mary and Huw feeding assorted fowls.

The Goose family out on the pond.

The goose-house was in the Home Field, a long way from the pond, and it was here, traditionally, that Mrs Goose had her nest. Once she had started brooding, she only left her eggs for very short periods to feed and defecate. There was no way of knowing precisely when she would do this for she did not necessarily get off to order when food was brought. We tried to keep watch so that someone could look at the nest while she was off it, but unless someone could be there the entire day it was almost impossible to catch her. And so we rarely did.

In a proper goose community, nests would be built beside a lake. Ganders would be nearby to guard the eggs while brooding females got into the water for a break. Returning to the nest, a goose would turn her eggs, dripping water on them as she did so, then settling her wet feathers over them. When geese are brooding in a house they do not always manage to turn all their eggs regularly. Perhaps the instinct is diminished without the other rituals of swimming and wetting. Or perhaps there is always some natural wastage in the wild.

Our Mrs Goose did not have access to water, the pond being too far away, and so her eggs needed to be wetted and turned by hand. The farmer's wife's hand!

Old Mrs Jones made one of her very rare visits to the farm to instruct Madge in the mysteries of goose lore. She was a tiny woman, hardly weightier than a well-grown goose.

'Quite easy it is,' said Mrs Jones, describing what had to be done in phrases translated word by word from the Welsh, which I will not corrupt by trying to reproduce in writing. She proceeded to show just how easy it was by walking right into the shed where Mrs Goose was sitting on her nest. With absolute authority she turned the hissing head aside and lifted the bird up. Keeping her tucked under one arm she splashed warm water on the eggs, turned them over and splashed them again. She put Mrs Goose back and returned outside. No doubt about it. It was dead easy.

When the next wetting was due, Madge set off determinedly with food and water, quite confident she could manage, after Mrs Jones's demonstration.

'If Mrs Jones can do it, I can do it!'

To be extra sure there would be no problem, she intended to put Mrs Goose out in the field with some food, shutting the door while she attended to the nest.

Time went by and I hadn't seen her come back. I went into the Home Field and approached the goose house. I could hear Madge talking and assumed she was having a conversation with Mrs Goose about the egg business. I waited a moment, not wanting to poke my nose in, and Madge appeared, just stepping back from the door. I was about to ask if all was well, but before I could say anything she had disappeared inside again.

And now I could hear what she was saying. She was repeating the same thing over and over, sometimes forcefully, sometimes despairingly:

'If Mrs Jones can do it – I can do it. If Mrs Jones can do it – I CAN DO IT.'

But she couldn't! And neither could I.

In the end, David did it.

And the next year she saved all the goose eggs and gave them out to broody hens. There was no problem about lifting these off while she watered and turned the eggs. The poor hens could only cover three or possibly four goose eggs and they must have been so uncomfortable, poor things, squatting on such huge eggs, their feathers permanently wet. They managed very well but were quite tattered and threadbare by the time the job was done. Did they notice, I wonder, how few and strange their children were compared with those of their neighbours? Did scratchy friends come up and comment on the size of their children's feet and beaks and say, never mind, I expect they'll grow out of it?

When the goslings began to feather there were more problems. They would be forever playing about in puddles, flapping those big ugly feet up and down and sticking their great shovel beaks under the water. It was time for them to join their own kind on the pond. We took them all down, watching over the little birds for a few days while they got used to the water and their new relations. We had to watch too while the big geese got used to them. These didn't recognise the newcomers as their own kind. They were intruders to be chased away, liable to be viciously pecked, possibly drowned. The foster mothers ran frantically round the edge of the pond, calling to them. They clucked 'danger' and 'food' but to no effect. The fledglings were in their element and there was no going back. They found their pecking order in the flock and learnt to dodge the big bullies and get a share of the food. As for the poor hens, they shook their heads over the ingratitude of kids and went back on the production line.

One day in 1948, the baker brought out of his van an extra package. It was a tiny white terrier wanting a home. He thought I would like her. I did. Very much. Nobody objected so she stayed and became my dog. I called her Pippa and took her everywhere with me when I went walking. I even carried her under one arm while doing the precarious climb round the cliff face from the second to the third bay in Llangrannog. She was a timid little animal and never grew very big. Remarks were made to me

Early form of chick rearing.

occasionally, some more direct than others, about why I wasn't keeping company with a boy instead of a dog; or it should be a baby I was carrying, not a dog. I smiled and said a dog was less trouble.

It was some months into the next year, when I was in the house and thought I heard Madge calling. There was something strange going on, an odd noise I couldn't make out. I hurried outside and saw Madge running down the yard towards me in great distress. She was crying.

'Mrs Goose,' she sobbed. 'The dogs have killed Mrs Goose.'

It was a terrible business. Not only for poor Mrs Goose and Madge, but for the dogs. The three of them had ganged up together, hounding the goose into a corner and attacking from so many angles she couldn't defend herself.

The accepted wisdom of the day was that once a dog had tasted blood it would become a compulsive hunter and killer and could never be trusted again. There was no cure and no recourse but to kill the dog. No one argued. Even while living in town I had heard this rule and now expected no other response from Cecil. It was awful but it had to be done.

Beautiful, timid Fan, who was a pedigree red and white collie; the streaky black and white collie, Belle, with her one brown and one blue eye; and my darling little Pippa. All were shot.

We never had proper working dogs again. I was given an unwanted collie pup from a farm along the way, but he caught distemper after only two or three months. The vet could do nothing for him but shoot him. Later on Madge got a stray corgi and then two beautiful pedigree corgi pups. They were jolly little dogs and stayed part of the family for a great many years. But they were house dogs.

Madge stopped hatching geese. Occasional years she bought in day-old goslings and reared them for Christmas, but mostly we stuck to cockerels and a few ducks. Geese were terrible things to pluck anyway and never seemed worth the effort with their big frames and small amount of meat.

There were a great many things Mrs Jones could do that Madge couldn't, but there were also many things Madge could do that Mrs Jones couldn't, so it evened out.

Chapter 14
Crops

THERE were Government Regulations about crops. They told us what we had to grow and how much to grow. These regulations were in force during the war and for two or three years after. Often they were sent out without any sensible reference to local conditions of land and weather. We were told to grow so many acres of wheat one year. The Welsh climate is utterly unsuitable for wheat, which needs plenty of sun and little rain in its crucial ripening period, conditions notoriously absent from summers and autumns in Wales. Without them the grain cannot harden and is useless for anything but chicken feed. It was a happy outcome for us in this particular case because chicken feed, like everything else, was scarce and rationed and our useless wheat quickly found a ready and lucrative market. Another order was for a water meadow to be ploughed up and potatoes grown there. Not surprisingly, most of them rotted in the ground. Still, since the system overall produced enough food to keep us all healthy through the war, it must have been mostly right.

The first crop of the year to be sown was potatoes. The ideal recommended time for sowing these was the third week of March. This allowed optimum growing time with the least likelihood of frost. It must have been a late sowing that first year, waiting for the snow to clear. I don't remember the occasion, though I surely would have helped. All

Sowing swedes and fodder beet. Cousin Doris is leading Bess and Henry is steering the beet drill.

cultivation and sowing must have been late, but harvests were good because snow helps to set nitrogen in the soil and the months of sunshine made for good growing conditions.

The potato field, or section of field, for we never grew more than two or three acres, was ploughed as soon as the land was clear of other crops and dry enough to be worked. If possible, ploughing would be done in the autumn. Then in spring, as soon as the ground was dry enough to work, it was harrowed, ridges would be drawn out and then muck was carted and spread in the trenches.

An intriguing thing when land was being cultivated, and especially during ploughing, was the way the seagulls appeared, first one circling overhead, then two in the furrow behind the plough, then suddenly all the turned soil was alive with 20, 30 or more. It was always an excitement and a mystery to have them drop out of nowhere, filling the rows as they squabbled for the fresh-dug worms and beetles. When they were in front of the tractor, they seemed to go right under the wheels before hopping to safety at the last moment.

There were no mechanised implements for digging and loading muck. It was forked out of the big heap into the cart, off the cart into the field, and then spread in the field, all by fork and body power. Small heaps of muck were pushed off the back of the cart into the rows, the vehicle going up and down the field until all the furrows were charged with their little steaming piles. The tractor and trailer ran with their wheels in two furrows, spanning the third that had the muck in. With the horse and cart, the horse walked in the centre furrow. If there were spare hands about, these could follow behind the carts and start spreading the muck along the trenches. Otherwise those driving and pushing it off would have to come back later to do the spreading as well. A great many trips had to be made back to the yard for refills before a field was covered. Completing the job could take two or three days. Today on a farm like ours, a hydraulic fore-loader and a mechanical muck-spreader would clear a muck heap and spread its contents on three or four fields within one day. Barring breakdowns.

The picking and planting of potatoes were partly communal jobs, but as all growers were trying to catch the best time and weather, there were not always spare hands available on the farms for help with planting. Non-farming people from round about, mostly women, could always be hired by the day to help. Sacks of seed would be dumped at strategic points in the planting area, from which we filled our buckets and went plodding up and down the rows setting the spuds in place.

Moulding-up is another of those very simple operations that are very difficult to describe clearly and succinctly. Well, for the likes of me they are. An ordinary plough cuts, digs and turns the soil with a disc beside a mouldboard, which has a sharp-pointed

Madge, Timmy and Cecil sowing cattle beans on a very cold day.

ploughshare on its nose. It turns the furrows and lays them flat on top of each other. The ridge plough has a sharp flat nose in front of two back to back mouldboards. It divides soft soil, turning half to each side to form ridges.

The old ridger was a very elegant implement, beautifully shaped and balanced as were so many of the old hand and horse tools. The handles curved up and over into a long downward sweep to the plough body. Here they merged together and curved over the moulding boards. There was a ring at the front to which would be fixed the whipple-tree with the horse traces attached. The flat nose on the plough was designed to dig in and lift the soil which the moulding boards then divided and pushed up to each side, covering the potatoes and shaping two half ridges. So you and horse and plough walked between two rows, making half a ridge on each side of you. You then, obviously, had to work down the next row to build up the other half of the ridge. As with all such up and down operations, rather than make complicated turns to drive back down adjacent rows, you would pull forward and make an easier turn into a marked row farther along the block. At the end of that row, you turned back to the one next to the row you'd previously been along and did the same thing again, completing one ridge on the right and building another half on the left. Eventually you have worked along every row, producing a field full of smart ridges.

Potato lifting was a more social affair altogether than planting. Time and weather for doing it were not so limited, allowing workers to be exchanged between farms. And whereas you had a whole row to yourself to plant, one spud at a time, gathering was done with a partner and in a crowd, there being considerably more spuds to be picked up and carried than there had been to plant. Which of course is the whole point of the operation.

Digging was done by a cunning machine consisting of a scoop that dug into the ridge, opening it up and leaving the soil and potatoes in a loose heap as it moved forward. Behind the scoop was a big wheel set at right angles across the ridge. This spun around, flicking the potatoes away from the soil and out to one side with little two-pronged forks on the ends of its spokes. A seat was set above the gearbox between the wheels of the digger. The driver steered the horse from here and lifted the spinner out of the ground as it reached the end of a row and turned to enter a new one. There he lowered it again. If a tractor was pulling the digger, a passenger was still needed to do this lifting business.

At my first potato-lifting, I was charging along full-blast, showing off as usual, though for myself as much as anything, intent on keeping up with the men. It seemed to me they were just loafing about half the time and yet I couldn't fill my pail and empty it as often as they did.

There were three farm boys in particular who were keeping very much together, all cloncing away as they worked. Sometimes as I passed they would be speaking English. I gathered they were discussing some farming business from the occasional words I heard. I was surprised they weren't speaking Welsh among themselves and supposed it was because Robert was originally from England. As I was working right beside them for a time they asked how I was getting on, did I like living here, was I walking out with anyone. They put their heads together and did boys' equivalent of giggling as they said this. When I looked round from emptying my bucket they were standing watching me but quickly turned and got on with the job. They had picked their way several yards ahead of me by the time I got back. I still heard bits of their talk. They seemed to be describing the merits of a working horse: Strong firm legs and straight back, a bright eye, good chest and lungs, plenty of pulling power. You'd have a bargain there. And for all their chatting and apparent lack of effort, they had still each emptied a bucket – I watched them go! – by the time I straightened creakily upright to empty mine. As I walked along the row with it I heard, 'Good childbearing hips and tits too,' then whistles and growls of laughter.

Potatoes were considered a land-cleaning crop because the amount of soil moving that went into their cultivation disturbed the weeds, killing many of them. They were also essential in the crop rotation cycle, still an intrinsic part of farming practice at this time. Corn would be planted in the field the year after a crop of potatoes and beets had

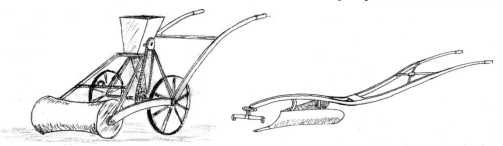

Left: Beet drill – a curved roller shaped to ride on a ridge. The wheel axle turns the cogwheel, driving the vertical chain. This turns a shaft in the seed box, having small, brush bristles on it which stir the seeds, stopping them from clogging together, also sweeping a few at a time into the planting tube. This scrapes a shallow channel on the ridge into which the seeds drop. A short length of chain joins the horse's whipple-tree to the loop above the roller. *Right:* Ridge plough. A very elegant implement. The long flat nose in front digs and lifts the soil. The moulding boards push through behind it, dividing and shaping the soil into half ridges. This first makes the trenches for planting the potatoes in and later goes along covering and building ridges over them. The whipple-tree is linked on the front, to which the horse's traces attach.

grown there, the heavy mucking of the roots continuing to feed corn crops for two or sometimes three successive years after this. In the last corn-growing year, the crop was undersown with grass seed, which it sheltered and helped establish to make a lush hay crop for the following year. Plenty of clover was included in the seed mix and this set nitrogen in the soil from its root nodules. And so that field gradually returned to grazing pasture and after a few years, the cycle started again.

The time for sowing corn and grass, whether separately or together, was ideally mid-April. Sometimes it could be done earlier, sometimes it had to be later. Probably most farms of any size (100 acres was a big farm in this area at that time!) would have owned a corn drill. Others could hire one from the War-Ag. Small farmers growing only one to three acres of corn walked their land, broadcasting the seed from a bucket.

The traditional local grains were barley and oats. Cardiganshire had been a big producer of barley for brewing before the war and was also well known for horse breeding, so oats were grown for the horses. When dairy farming became profitable, the land was mostly put down to grass with enough barley and oats being grown to feed cows and horses in the winter and provide straw for bedding. Quite a lot of oats were still grown on their own for feeding to horses, but barley was mostly grown mixed with oats. This was because barley was very short stalked and the taller oats helped to keep the heads from drooping on the ground. The two grains ground together were good feed for cows. Cecil also grew cattle beans, not a locally grown crop, to supplement the grains. They were ugly, black stalked things when ripe, hard and crackly, but beautiful to see and smell when in flower.

Grass seed, whether for undersowing or grown as a crop on its own, was sown with a fiddle. This was worn like a rucksack on one's front and comprised a sack containing the seed, a box to feed it through, and a bow to work the mechanism. An adjustable slide opening at the bottom of the box was set to allow the required amount of seed to feed out onto a metal disc on a spindle. The disc was spun round one way and then the other by pushing a bow from side to side, like playing a fiddle. As the sower walked across the field doing this rhythmical movement, the seed was spun out in an arc in front of him. The more regular the rhythm, the more regular the distribution and growth of the seeds.

The last of the spring sowings was of the beets and the kale, all of them very susceptible to frost. The time for roots was early May, possibly late April in a really mild year. The seeds were sown on the tops of ridges which were drawn up in the same way as potato ridges, only without digging the drills first, of course. The seed was sown with a turnip drill. This had a barrel-shaped wheel like a roller, but concave to run along the top of the ridge and curving over at each side to hold it in place. A funnel-shaped container held the seed and from it a long spout stretched down behind the roller, planting seeds just below the surface on the flattened top of the ridge. The person in charge followed behind holding the curved handles and the horse's reins. This method of sowing allowed the seeds to be placed at fairly precise intervals and weeding to be done between the rows.

Swedes, mangolds and fodder-beet were the three roots grown, though not necessarily all of them every year. Like a great many of these old basic crops, they were a lot of trouble for what seemed little return, though in fact the contribution they made to health and comfort of animals in those days was quite important. Their main function was to provide liquid to beasts whose sole diet otherwise was dried grass, straw and dry corn meal. They also provided fibre, vitamins and minerals and what was probably just as important as all of these, interest and variety. Most cows and horses were tied in sheds all winter, through November and March, with no water except what might be carried to them in buckets, and no exercise unless they were lucky enough to be let out for limited access to pond or stream. Very boring.

Mangel-wurzels (mangolds) were the main cattle beet. They were put through a root

Gulls suddenly turn up from nowhere to follow the plough.

Win and Charles, very close friends of the farm and family.

pulper and the bits and the juice were scooped out into the cows' glazed mangers. Swedes were mostly for human consumption. As they survived frosts reasonably well, they stayed in the fields much later than mangolds. Fodder-beet was a new crop. It wasn't grown much because roots as a crop for cows were already dropping out of favour anyway.

Beets were harvested in some of the worst weather of the year, in November. There was little else to do on the farm and it was one crop that didn't mind rain, snow or frost on it during its harvest; unlike those doing the work! The job was reasonably warm once you got into the swing of it; and swing was the operative word. You swung your body down, pulled the root and swung it into the air by its leaves. In the same movement you were standing upright, swinging the beet so it was stiff enough to cut the root from the leaves. The timing and arc of swing had to be just right to make the severed beet continue its flight straight into the trailer creeping along beside you. As the root flew forward, the leaves were slung backwards, and you swung down to grab the next plant.

Some people stored their roots, the potatoes and the mangolds, in outside clamps

Showing off a seagull chick from a cliff nest.

heavily covered with straw to protect them from frost. We occasionally had an outside clamp, but mostly there was room enough to heap ours up in the big barn, though they still needed some straw covering.

Kale was mostly sown the same way as the roots, with a turnip drill, though sometimes it was broadcast. It was sown in May. When we first started growing it, it was customary to take a cart to the field and cut a load with a sickle or billhook, tipping it into the cows' mangers to supplement their other winter feed. It was a horrible job on wet days. We tried letting the cows in to help themselves but they tramped all over the field, breaking the plants down and wasting them in the mud. Not only that but they ate off all the choicest leaves in the first few days and then said they couldn't be bothered to eat the stalks on their own.

Later we got electric fences and set them up to ration a day's supply at a time. It worked moderately well but in the end was possibly only slightly less trouble than cutting and carting. The problem was with the electricity shorting. Anything touching the wire earthed it, taking away the sting. No matter how carefully posts were set in and the wire tightened and surrounding plants cut away so nothing could possibly touch the wire, without fail the cows would find a way to beat the system. If the fence was working well with a strong current going through it, none of them would lean across it to reach distant stalks, but when the battery started to run down, a brave cow would soon find she could push hard on the wire without much discomfort. Other cows would soon be doing the same, which reduced the strength of the shock still further. Before long the fence went flat on the ground and the whole herd stepped over it, or worse, caught their feet in it, tangled and broke it. They might also push insulated posts over with their heads, or more clever still, they learnt that their horns were non-conductors and they would walk along stretching the wire far beyond the day's ration by fixing a horn on it. They cut deep grooves in their horns doing this. One did actually cut a tip right off once. Horns are like our fingernails, they have dead ends but contain nerves and blood lower down.

They were not all clever enough or determined enough to work these things out for themselves, but cows are great watchers and followers and it only needs one to lead. Once one was through it was up heels and full charge for everybody. They went through at least once or twice a week, sometimes two or three times a day. And the days they liked to choose if they possibly could were the wet ones.

Trying to get a herd of cows out of a field of wet kale was absolutely no fun. These are plants about six feet high with big floppy cabbage leaves all the way up their stalks. But though it's no fun for humans, it's absolutely hilarious for cows. There are few things a cow likes better than a good chase-about when she knows she has the upper hand. And in a field of kale she had the upper everything. It was catch-as-catch-can in a swampy jungle. Here were 15 or 20 huge lumbering animals in a small field and you couldn't see any of them. Chasing them was no use anyway. It needed a line of beaters. Still we had to try. Even when there were two of us working together, it took an awful lot of shouting and chasing backwards and forwards before we got them all together and out. Gathering three or four and thinking you'd get them out the gate and come back for the rest was no good. They strung us along for a bit, huddling for a moment in the gateway, then someone said, right-o, girls, that's enough, and it was heads down and heels up and they were gone off in all directions except through the gate. The times I have been reduced to a screaming, slobbering, gibbering maniac by trying to make a herd of cows go somewhere they didn't want to, I'm surprised I have any blood pressure left to take.

I think we did not grow kale for more than three or four years.

Chapter 15
Paying Guests

BESIDES the farmer's wife's traditional perks of money for eggs and various trussed fowls, Madge also took in paying guests during the summer. Thinking back now, I cannot imagine there was much actual profit in the business. Certainly these people paid, and certainly they would have paid the going rate for that sort of holiday. Except that it wasn't 'that sort of holiday.' This holiday had huge meals of very good quality food and also considerable amounts of alcohol. I think we always had at least one gin or whiskey a day, and as all visitors lived in with the family, they had the same. Spirits were still very hard to come by for two or three years after the war, but Madge and Cecil had old customer connections with big London stores who supplied some; and friends always brought bottles with them.

There were special arrangements for the closest of friends and 'family'. Emmy and Felicité, like any family, came more or less any time they liked, joining in the work and fun, helping out in emergencies and being looked after in their turn. They brought food and drink contributions and certainly didn't pay as such. Hilda was in much the same category. Very close friends, Stanley and Nellie, Win and Charles, insisted on paying for annual holidays but came just as friends other times. All of them brought luxury town goods to contribute to the general victuals and other entertainment.

I remember a Christmas when Stanley and Nellie arrived just as we were about to set out across the fields for the festive gathering of holly and ivy and branches of fir trees. They wanted to come and join in the fun so dumped their bags down in the hall while we all stomped off over the fields. Coming back, we put most of the greenery down outside the house and I went away inside somewhere – getting cups of tea ready probably – while they sorted out luggage. I went back out after some minutes and was surprised to find everybody still milling about in the hall and porch, seeming to be looking for something. I had been looking forward to the bags being unpacked by now. I wanted to see what luxury goodies Stanley and Nellie had brought.

'Bluebelle!' said Madge to me, 'come and see if you notice anything strange.'

I walked across, went backwards and forwards and in and out as they'd been doing, but I couldn't see what she might be on about.

'Do you smell anything?'

The farm had its wafts of this and that, but yes, now she asked the specific question, I certainly could smell something potent close among us. I sniffed round the boots and the greenery. Nothing untoward there. I turned and went back into the hall.

'What is it?' they were all saying to each other.

I wrinkled my nose and my brain, moving a little closer in towards the luggage.

'It's cheese,' I said.

One of our Christmas treats! Funny how much nicer it smelt once we knew what it was.

Most of the paying guests were slightly more distant friends or old work colleagues. They all accepted the strange living arrangements with remarkable equanimity and returned year after year, becoming closer friends as they did so. Some of their children not only remained friends but even moved to the area permanently.

None of these more distant friends were exposed to the nudist aspects of our life.

To leave us with freedom and space for our morning rush around, Madge and I organised breakfasts in bed for them all. I took up trays with pots of tea, grapefruits, toast and boiled eggs, delicate dishes of marmalade and curls of butter. Some people had bacon and eggs. After a suitable time I took away the empties and delivered jugs of hot water for washing. One couple had the strange habit of from time to time being on opposite sides of the bed to those I expected them to be. It was explained to me that in the night Mary kept edging farther and farther across the bed until her husband found it easiest to just get out, walk round the bed and get back in the other side.

There must have been a number of people who came only once for a holiday, but they have faded from memory.

Except for three.

Early on there came a colleague of an old friend of Madge's. She was in the Civil Service, as Madge had been, and was what used to be called a Maiden Lady. There were subtle differences between maiden ladies, spinsters and old maids, which were not necessarily to do with age or position in society. Vera – I do remember her other name too – was a lively, outdoor person, kind and friendly. She didn't seem particularly prim and proper. She might have been fortyish, or she might have been thirty or fifty. I took her breakfast and her jug of hot water for washing and she was pleasant and appreciative. I remember she was always knitting. I can see her now on the beach at Penbryn. I was splashing about in the sea and she was walking along the beach knitting. She was making a white two-piece bathing costume for me. It was really smart and fitted perfectly. It never did that awful stretching that hand-knitted bathing costumes were prone to. She put red trimming round all the edges and when I wore it the red ran out in long, embarrassing streaks down my legs. The whole thing remained stained with red splodges ever after. I wore it anyway. Then one day when I was swimming, I went down under a great breaker and came up without the top. I spent some time prowling up and down searching the waves for it, to the amusement of a lone man on the beach. I didn't find it.

The lady seemed to be quite settled with us. We were so used to our way of life and to other people accepting it that we didn't give any thought to the possibility of unresolvable problems.

One day Madge went into the dairy and called to me to bring a cloth as someone had spilt something on the floor.

'Did you drop something?'

'No.' I said.

'It's funny. There was a mess just there yesterday.'

I looked up. The ceiling was wet.

I squawked and ran. In the Elsanole immediately overhead I kicked the pee pail – gently! It was empty. I lifted the lid of the Elsan. Even in the gloom of that darkest of corners I could see the fearsome contents level with the seat and seeping away somewhere underneath it. That poor woman! Unable to bear the indignity of squatting on a tinkling pail, or of going outside, she didn't know what else to do but continue to sit on the nearest thing to a civilised lavatory available. She didn't dare to think about what was going on underneath her.

Of course it should have been checked and wiped around everyday, but we didn't bother with such excessive fussiness. The thing had been checked and cleaned when Vera arrived and it should have been all right for one person for a fortnight. I did check the pee pail. This would have been emptied every day. As it was unused, Madge and I assumed that she was using the hedgerows. We were impressed by such unexpected freedom from inhibition – mistakenly and unwisely, as it turned out!

I was very good at waiting on people. It earned me the pet name of Flunkey from Flicit, and she still calls me that. It also earned me some occasional tips. Or perhaps they were left for less personal reasons. I must have been helpful and nice to Vera though, for she left a note as well, thanking me for 'more things than I can mention.'

I can't remember who the next awkward lot were; only that the son was named David. I don't know how or why they came, except that they must have been friends of friends of friends. I would remember whose friends they were had they been friends only once removed. It's an interesting fact that whereas friends of friends always settled down happily, friends of friends of friends never did. It wasn't just the way of life, there was never the rapport between us and them either. Luckily there weren't many of them. We were always full with people we knew and liked anyway.

It must have been 1948 when that second disastrous lot came: David and his mother and father. They were already not speaking to each other when they arrived. Mother was furious and son was sulking. I can't remember Dad at all. Mother was ranting on about David's stupid obsession with trains. On the drive down, they had stopped in a small town for tea and David had disappeared. He was about 19, by the way. She quickly realised there was a railway station across the road and sent Dad to drag him back. After about 10 minutes Dad came back having been unable to find him. Mother went off and also failed to find him. It seems he'd met up with a friendly railway worker who'd taken him to look over an engine standing in the siding. Half an hour or more they'd had to wait for him, Mother fumed. They spent most of their holiday fuming and sulking.

Then one day, I don't know how he managed it, the parents went out without him. He seemed happy for once, wandering round the yard looking at things. He was even talking to me a bit, which he never normally did, and he kept asking if he could help. I was mucking out the calf pens and told him he could get some straw and shake it out for bedding in the shed I'd just cleaned.

'Better put an overall on,' I said. 'There's one in the porch.'

He did the bedding and put hay in the manger. Then he got a bucket of water for the calves. He looked remarkably normal and happy. He stood about watching me again and asked if he could have a go at emptying the barrow. The heap was not very high yet. There were two planks to run up but no corners to go round. If he had been one of the pushy, show-off types who come elbowing in with an, 'I'll do that for you', I might have stood back and hoped he'd fall in. But he wasn't. He was a nice quiet young man who badly wanted to join in a life away from the swamp of his family. I tried to put him off, telling him it was not as easy as it looked, but he said he had watched how I did it and couldn't he please have a try.

Well, what the hell! A bit of muck never hurt anyone anyway. I paddled in it all the time. I told him he'd be all right if he aimed the wheel straight at the middle of the plank and kept running. I said if the barrow slowed down he wouldn't be able to control it.

'Don't try to keep going if it gets too heavy, and don't try to stand and hold it. Just give a hard push and a twist to the handles and tip it over the side.'

Sure enough, as I'd feared, he hadn't the confidence, or the strength, to keep running. But he did remember my warning. As the barrow slowed down he gave a great heave and tipped it away from him off the plank. Unfortunately he failed to let go of the handles until both he and the barrow were well past the point of no return. He flapped his arms in a wild attempt to regain his balance then sprawled full length in the soft, wet, lower reaches of the heap.

I was absolutely horrified and choking with laughter.

He dragged himself upright as I ran down the yard to help him. The poor lad was dripping with liquid manure and had more solid stuff and straw stuck in his jersey and hair. He might have been crying.

'Whatever will Mum say? Ohhh.' He shook his hands and wiped his face distractedly. 'Oh my jersey! Ohhh. She just made that for me to go to university. Ohhhh.' He stood and wailed, much more worried about his mother's reaction than the muck itself. Not surprisingly. She scolded him all the time, publicly and aggressively. I didn't know why he was at home now without her.

There wasn't much I could do. Mod cons like a bath and washing machine would have set the worst of it to rights but all I could offer was the tin bath in the relative seclusion of the engine shed behind the cow shed. I took him soap and towel, buckets of water, and an overall to cover him afterwards while he went indoors for clean clothes. I told him to put his clothes in the dirty water when he'd washed. He could stamp them about a bit; then tip them out and rinse them under the cold tap before getting back in the bath with them and tipping the rinsing water over himself.

'Stamp them about again while you do it,' I yelled.

And of course his mum and dad had to come home just as he was walking up the yard in the overall. It was sadly, embarrassingly, hilarious. I never saw or heard of them again.

However, immediately after they left there came another man and wife and son. Madge and I were waiting with some trepidation on the Saturday afternoon, for these were friends of friends we had not met before. We knew their names and ominously the son's name was David. I can see quite clearly, in that Wordsworthian eye, Madge and me standing by the sitting room window, looking out at the little car that pulled up on the far side of the yard. It was a lovely sunny day. We watched as the two men got out, and we looked at each other and made a face! They were weedy and they wore glasses. The woman came round and we thought our worst fears justified. She seemed efficient and bustling.

'I'll go and see what they're like,' Madge said. 'You wait there.' I stood and watched while she walked across and greeted them. After a moment she turned and beamed at me. I went to get on with the tea.

They turned out to be kind and cheery and interesting to talk with. And they got on well together. They became our great friends, right through to the third generation, and some of them moved into West Wales. From David I learnt a bit about birds, for they were an interest of his. As were trains, but luckily that was as close as similarity to the other David and his family came. We walked down to the woods in the evening to listen for nightjars. He was sure there should be some about. We sat and listened.

'There!' he said.

I shook my head. 'I can't hear anything.'

'It's gone. Listen.'

I listened. Nothing. Not a tweet.

'All I can hear is a motor bike.'

'That's the nightjar.'

I listened to it brr-uh-urrring away. It still sounded like a two-stroke motor bike, but I'm glad I heard it for they disappeared from the district in the next few years.

Len and Jessie and David came as friends of Stanley and Nellie. In their turn, they introduced friends of their own. Again, friends of friends of friends. These were our third incompatible family: a man and woman and small daughter. They came only for a trial weekend, and this proved more than enough for all of us. Except that, strangely, the little girl came back years later with her husband and children and they have all become our good friends. But they didn't stay in the old farmhouse. They went to Madge's newly built house with all mod cons.

Other regulars were a Welsh family from Bridgend. Llew and Mary and their son, Huw. They came together with Mary's sister, Ethel, and her husband Les and daughter

Susan. All loveable people, and great fun in a quiet way. They were also musical. Les played my old piano and we had long, enjoyable evenings of music. Sometimes Ethel sang. After the first year, the children stayed on for all of their summer holidays, after their parents went home. I think Susan was nearly 10, Huw 11. They loved being on the farm and we enjoyed having them. I remember Susan dragging me out of the tent before morning milking that first year to go with her to see the horses in the field. We still had the young colt and filly, Punch and Judy, who were quite wild and frisky. Punch especially was inclined to rear up on his hind legs and was really frightening. Susan was hiding behind me, but I was as terrified as she was. She is another of those who has moved into Ceredigion to live.

So many of the pictures I keep in my mind seem quite trivial, but they were important parts of the everyday companionship of my life, most of which remains very clear and alive. I have forgotten, or never knew, the intricacies of the merely functional things to do with measurements, crops and machines. I have to check details of those with Patrick. I couldn't describe the parts of Bess's harness, for example, and how they fitted together, things I dealt with automatically every day, but I bet I could go and do the job now without thinking. All the important bits, though, the people and their doings and sayings, these are always in my head and heart.

Chapter 16

Haymaking

THERE are no hay meadows any more. No tall mixed grasses with shapely seed heads and lovely names like cocksfoot and Timothy. No wild grasses like Yorkshire fog and barley grass, or wild flowers like marguerites, scabious, corn marigolds and poppies. You never see thick deep tresses of grass, full of lilting colours, chasing endlessly from hedge to hedge in a summer wind like deep ocean waves; nor fields of knotted grass tipis twined together after wild dervish dances in stormy whirlwinds.

Grasses are different now. They are bred to grow low and lush and seedless.

My experiences of haymaking started with the basic pitchfork and horse and cart, though all of the mowing and some of the carting was done using the tractor. The range of haymaking equipment that came with the farm was all built for natural horse-power – as were all other implements, of course. Very little machinery was actually custom built for tractors at that time and as tractors began to take over the fieldwork from horses, so they also took over their machinery. With no hydraulic lifting systems, what would work by being pulled behind a horse could equally well work being pulled behind a tractor. Sometimes metal hitches were made by the local blacksmith to attach things, sometimes the shafts were simply tied to the tractor hitch, or whatever else was convenient for the purpose.

And they were so clever, these machines. The mower didn't just cut the grass and drop it flat, it had a cunning little blade that slid under the stalks and rolled them sideways to leave a space of about 12 inches between the cut swathe and the standing grass. This left a clear strip for the horse or tractor to move along on the next round without having to run over the crop. For the same reason the outside edge of the field was first cut with a scythe to 'open' it.

The grass could be cut wet. Indeed, it cut more easily wet and so mowing would be done late in the evening or in the early morning. You drove clockwise round the field with the mower blade on your right. The cutter-bar with the knife in was far enough back for the driver to be able to watch and see it kept working properly. By some magical system of shafts and gears, the forward movement of the mower wheels caused the knife to move from side to side in its casing. It actually cut with a scissor motion, the grass being trapped and sliced between the blades of the knife and the hard metal fingers on the casing. The forward movement alone was too slow to slice through the stalks. The modern disc mower moves so fast it slices through grass before it has time to duck.

Great skills are required for haymaking, as with so many farming jobs. Not least of them is judging the growing crop for its best cutting time and the weather for its best promise. Working with pulled machinery also needed special skill and judgement. When mowing, for example, you did not simply turn corners. At each turning point in your rectangle you had to drive forward over the rows of cut grass, at the same moment pulling on a piece of cord to raise the cutter bar. The tractor had to be driven in a wide enough circle to take the length of your trailing implement without dragging the laid grass, ending up at right angles to your last cut row. Without losing speed, you lined up for your new cut, pulling the cord to lower the blade at the precise moment when it

Horse and cart almost submerged under this first load of hay.

1970: a load ready for home. These are early bales, a comfortable size to handle but no good for rolling in.

would meet the standing grass. Failure to lower in time meant a few stalks would be left standing. Nothing desperate. Lowered too soon, however, or lifted too late, the knife is left whizzing backwards and forwards through a pile of loose grass, winding it and binding it round the blades. You lose time and energy clearing it and you make more work for the person going round the field shaking out lumps with a fork. Hydraulics eliminated all of these problems, for you could raise or lower your implement at a touch and you could reverse to change position. No way can horse-drawn implements be pushed backwards.

The day after cutting, and weather permitting, the hay turner was taken along the rows. This not only turned the underneaths up to the sun and wind but also moved the grass onto a drier piece of ground. It was repeated on the third day. Turning was done at about 11am when the dew had dried. With a good drying wind and some sunshine, a perfect crop might be rowed up, collected and stacked on the fourth day.

Loose hay from the fields was loaded into carts and trailers and unloaded in the stackyard into haystacks. We had an elevator that was attached behind the trailer and

Early silage system: fresh grass, spread and rolled to remove air, cooks in a clamp. Cows help themselves to their daily ration in winter. The barn alongside contains a similar clamp. On top are 900 (and 3!) bales of hay. Some will be eaten, others restacked leaving silage available for use after the outside clamp is finished. Assisting Venableses await next load.

this grabbed the hay from its row, trundled it up on its moving belt of mesh then tumbled it off and over at the top. One small person with a pitchfork stood there to receive it. With a big tractor-trailer there would be someone else at the far end taking some of the hay and together you had to build a well-balanced, stable load. The technique was to overlap layer upon layer from the edges inwards, thereby binding everything together and pulling the weight towards the middle. This whole process was carried on, the middle always being kept slightly lower than the edges, until the load was as high as possible without fear of it falling off on the bumpy ride back across the fields. Lying on the top for this slow plod or chug home was one of the joys of the job.

If you were loading a smaller cart on your own, you would sooner, rather than later, be overwhelmed by the avalanche descending non-stop upon you. With luck and careful observation the tractor driver would eventually become aware of your predicament and stop. Shouting was useless through all the machinery noise. My husband recalls the time his mother was alone on a trailer like this and was almost buried. He tells it as a very funny story but it could have been a ghastly accident. The only accident I ever had was on a very windy day, struggling to catch hay that was coming down diagonally from the loader, straight over the side of the cart. It was a bit like a mad inverted game of badminton where you're trying to catch these wild fluffy shuttlecocks and ram them down. As I rammed one forkful down onto the cart, I also stabbed the prong of the pitchfork right through my foot. Luckily it didn't seem to strike anything vital and I pulled it out and carried on with no ill effects.

It took so long, that early haymaking, and the weather was always wet at some stage, making it take even longer. We thought ourselves lucky if we got just one field of good quality hay in, having had no more than a quick shower on it. The more rain there was, the more time the grass had to lie about in the field and the more it deteriorated in quality. Every time it was turned over or shaken up to get air through it, more of its nutrients died and bits of the plant fell off. In desperation, hay was often taken in too soon, resulting in moulds growing. This made very poor quality, unappetising feed. At its worst, the stack would heat up and catch fire.

Clever new machinery – better, bigger, faster – hugely increased efficiency in harvesting grass and corn, and ensured they were in the best possible condition for storing and feeding. And the same for many other crops as well. This in itself would allow more animals to be kept on the same acreage and would also improve the quality of their milk and flesh.

But these improvements came much later with revolutionary changes to tractors and implements. The mere fact of a tractor moving faster than a horse did not mean that it would therefore move round a field mowing or turning the grass, or doing any harvesting, cultivating, or other field job more quickly. The implements themselves could only work at low speeds because of their simple structures. The only advantage was in pulling power. The tractor could pull a much heavier load than a horse, so bigger trailers with bigger loads meant crops came home that much more quickly. Still, even being able to drive from one place to another at 10 miles an hour instead of five was an immediate advantage, because all minutes saved were worthwhile when you were trying to finish a job before the bad weather set in, or daylight ended.

Most of our farm work was done with the help of both horse and tractor in those first few years, but Bess played a very important part in haymaking right to the end. Until baling came in. This important job was pulling the rope that worked the hay grab.

The bulk of the hay was stored in one huge stack under the high curved roof of the dutch barn. To get it up there we had a grab and pulley system. The trailer load of hay was backed under the barn and a great claw grab was lowered from the roof. The person on the load, quite often me, pulled the claws out to enclose as much hay as possible, then wedged them back in. A shout of ready went from me to someone at the end of the stack, who passed it on to someone way below holding onto Bess. Bess walked forward, pulling the rope that went from the grab, up and through a pulley in the roof, along the roof beam, out over a pulley at the end, down to the ground and through another pulley before being attached to Bess's whipple-tree. This is a bar slung behind her rump, attached by chains going forward on each side to clip on to her pulling collar. All implements were attached to the whipple-tree for pulling by one horse. A cart was not. That was pulled by chains from the collar along the shafts.

As Bess pulled on the rope, the grab rose up to the first pulley where it clamped into a little four-wheeled trolley, at the same time clicking shut a lock holding the claws together. The trolley travelled on along the beam, carrying the grabful of hay. When it reached the position where the workers were building the stack, one of them signalled to me and another shouted to the man with the horse.

These were signals for me to pull on the rope I was holding and for Bess to stop pulling on the rope she was holding.

A sharp, hard jerk was needed from me at this point to make the claws spring open and release their load. If the hay was slightly damp, a deal of force was required to persuade the metal grab to open and I often wrapped the rope round my hands and wrists to get a firm hold. Occasionally this caused me to be hauled into the air and left swinging because there had been a misunderstanding between Bess and her handler about whether she actually wanted to stop right there, right then. Or sometimes it could have been set up as one of those merry country jokes.

Not all the hay was unloaded using the grab. A lot was passed from load to stack with just pitchforks. Outside stacks had to be made in this way but there was also a side barn without a grab. Later Cecil had a new dutch barn added to the yard that did away with any freestanding stacks. Other than corn or straw stacks, that is.

We never seemed to have much help with haymaking in the first two or three years. Though thinking about it, it was only later when we had two tractors and two big

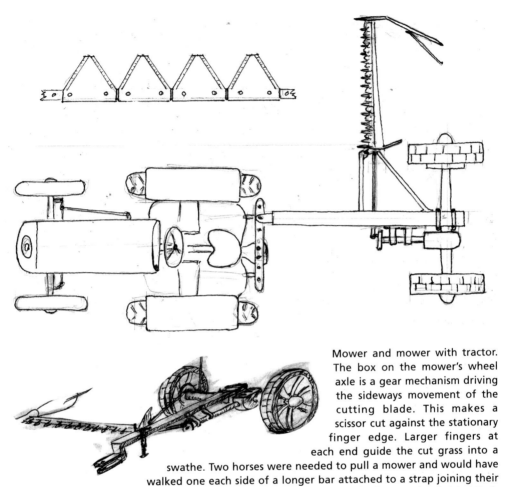

Mower and mower with tractor. The box on the mower's wheel axle is a gear mechanism driving the sideways movement of the cutting blade. This makes a scissor cut against the stationary finger edge. Larger fingers at each end guide the cut grass into a swathe. Two horses were needed to pull a mower and would have walked one each side of a longer bar attached to a strap joining their collars. The pull was through their traces to individual whipple-trees linked onto a bar behind their flanks attached to the 'tow'-bar. The whipple-tree bar swivelled to allow individual movement when turning corners.

trailers that there would have been work enough for the half-dozen or so extras who became our regular helpers. People on holiday often enjoyed doing bits and pieces and occasional young men from the village came up. I had some very pleasant times making hay with young men from the village. At one time during that first year, Madge said to me, 'The fellows are coming up from the Pentre this afternoon.'

Great, I thought. (Well no. It was probably smashing then!)

There were three very nice, handsome young men at the Pentre Arms who would be an asset to any farm job, but one especially was my favourite for haymaking. The job at that time involved a lot of lying about on soft fresh hay, waiting for the next load to come home. Qualified experts were always welcome to come and help out with this when I was working on the stack. Some more welcome than others. Some more expert than others.

Time went on and they didn't come.

Two friends of Madge's came later in the afternoon.

'Come and meet Paddy and Eric,' Madge called.

They seemed quite nice but they didn't do any work. I mentioned in a diffident sort of way that the Pentre boys hadn't turned up and Madge said she hadn't expected them.

'You said this morning they were coming.'

'No. I may have said Paddy and Eric were coming.'

'You said the fellows from the Pentre.'

'Well, they are the Fellows from the Pentre. Paddy and Eric Fellows. They're staying at the Pentre.'

I was greatly disappointed by this misunderstanding at the time, but Paddy and Eric were very nice to me and I had some lovely holidays with them in their home. I still keep in touch with their daughter, Veronica. Her son and grand-daughters continued the love-affair with Llangrannog that grabs so many people. Veronica reminded me they had a puppy, Ianto, from our bitch, Fan. She came visiting with him a year or so later and they were chased across a field by a mob of young cattle as she was gathering holly for Christmas. This can be a terrifying experience and I'm not surprised at her saying she and Ianto jumped a hedge to get away, though she added she went more through it than over. The creatures were probably only being curious, but a strange dog in the field will always excite animals. Even if you know them, it takes a lot of assurance to stand your ground and expect them to play by the rules and divide around you. They might knock you down simply by barging against each other or kicking their legs in the air. And once down, you would be very lucky not to be kicked or stepped on. Cows with young calves are the ones to be avoided. They can be really dangerous and more likely to attack you than most bulls you might meet. Flicit was once attacked in this way, not on our farm, while taking a cow and calf home from the field. The cow charged and knocked her down, then stood over her stamping on bits of her and swinging her head trying to stick her horns in. Flicit huddled low on the ground, covering her head. Luckily the cow's horns were the wrong shape to do much damage from that angle. Luckily too there was a man there to drive the beast off, but still she had two broken ribs and a lot of bruising.

Cecil made a point of doing no work on a Sunday unless it was absolutely essential. This was partly to ensure some regular rest time but mainly to avoid disturbing the peace of those who cherished Sunday as the Lord's Day. The exception to the rule was haymaking. Good hay was so difficult to get and so essential for winter feeding of stock that every dry moment had to be made use of. Many more years went by before local Welsh farmers would risk damnation in this way.

We had our own Bloody Sunday in one of those years when several good, chapel-going Welshmen did come to help us. Well, reasonably good. It was a lovely sunny day but our series of accidents would certainly have tried the apostasy of the weak minded. It culminated in Dai getting a finger caught in the grab, which was very nasty, and then a beam breaking and falling on Austin's head. He had to be taken to the doctor for stitches. A nurse in the village refused to treat Dai's finger, saying it was payment for his wickedness in working on a Sunday. She was treated to some very strong words from Dai's tough little wife, Bessie.

Small rectangular bales were starting to be made in the early 1950s and very soon there would be no loose hay except on a few smallholdings. These bales were a very useful and convenient method of making and feeding hay, relatively easy to handle. They weighed around half a hundredweight (25kg), and were built into big rectangular loads on tractor-trailers in the fields, and then packed carefully into the dutch barns in the stack yard. This was a hot and dusty job and took a lot of skill to build a secure stack, getting as many bales as possible into a limited amount of cover. By the time the modern monster round bales came along, virtually untouched by human hand from planting to feeding, I had happily retired from the business.

Most of those bales you see these days, tidily wrapped in various colours of plastic sheet (later to decorate road and riverside with its cheery streamers and unrotten heaps), is not actually hay, of course, but silage. The perfecting of silage-making techniques was one of the big gifts to farming of the 20th century. For your average, general-purpose farmer in Britain, that is to say. It did not at once make the physical work much lighter, nor was it any quicker in actual working hours than haymaking, but there was no longer the permanent, desperate worry of the weather to cope with. You can make silage when the grass is wet, whether inside itself from sap or outside from rain. Not too much of either, but considerably more than hay would stand.

Silage could not in any practical sense be made without the advanced machinery that came with bigger tractors and hydraulic-power driven implements. The weight and bulk of fresh-cut grass would need an army of slaves to move otherwise. Even the early silage-making method of tipping grass into a pit, spreading it by hand, then rolling with a tractor needed a certain amount of sophisticated lifting and tipping equipment. None of it was possible with horse-drawn implements.

Chapter 17

Our Modernest Conveniences

With every testimonial
And sacred ceremonial
Penrallt did thus experience
Her modernest convenience.

THUS (and onwards for several stanzas!) did I hail the switching-on of electricity from our generator. It was Easter 1951. The installation and wiring had hung about for a long time, partly, I suspect, because money came in bits and pieces but also because the buildings and terrain were particularly difficult to work with. Stone walls three feet thick do not lend themselves to easy drilling to get wires through and fixed to. But there it was at last.

Lights came on at the touch of a switch and maintained their strength without need of being pumped up every hour or so, and without low-hanging heavy metal cases for people to bash their heads against. There was nothing to carry from room to room. The light was there waiting. We had an electric cooker and a fridge and we could click on an electric fire for chilly moments. There was no longer the annoying, fiddly business of having to change the heavy accumulators that ran the wireless and ensuring that the empty one was taken up to Glyn's garage at Brynhoffnant for recharging and a charged one brought back before the working one died.

And joy of joys, we could use the radiogram. No more winding up the old gramophone, no more having to put on and turn over all the records one by one. On the radiogram there was the device allowing three or four records to be piled up together which would then drop and play, one after the other, on the turntable. Madge and Cecil already had several symphonies and concertos with the automatic coupling required for playing this way. The records were very thick and heavy. They dropped onto each other with a tremendous crash you'd think would smash the lot. They must have been remarkably tough.

There were drawbacks. They were slight though, compared with the benefits. The generator did produce the amount of power demanded of it, but it had to pause a moment and take a deep breath when demand was suddenly increased. So when you switched another light on, they all dimmed slightly before returning to full power. And the same with the radiogram. We would be drifting away in a dream of Beethoven or Brahms when the fridge or the oven broke in and the music droned into a slow growl before squeaking up to its proper speed again.

We had to be very careful to switch everything off at night before going to bed and had to remember not to switch lights on during the night otherwise the engine ran on. We got so used to the sound that sometimes we slept, or at least got into bed, without really noticing it. Then the noise gradually infiltrated into somebody's brain cells until they had to get up and turn off the offending power wastage. Unfortunate occasions did occur when certain of us came home late from courting in the wild woods or boozing in the Pentre Arms, crawled into bed and carelessly left a light on. The worst of these occasions was when a thermostat was involved, causing the engine to turn on and off intermittently. This happened one time when Madge and I had gone away for a few days. Cecil was tormented by the engine starting up in the night without any reason that he could find. He checked electric gadgets, traced leads and connections. Nothing. It was only when Madge and I came back he found out she had left her thermostatically controlled electric blanket switched on! They were sleeping in one of the huts so had separate bunks. Had they been in the house in their one bed, of course it would have made itself obvious immediately.

Only as I'm writing this do I realise that the fridge must have been turned off every night. There would have been nothing in it anyway that would not, in more primitive times, have had to survive in the open air.

Electricity in the house did make life very much easier, unlike those farm machinery improvements which actually enforced more and more work on farmers. Apart from the convenience and the time saved by not having to light, clean and generally wait about for the paraffin lamps, not to mention banging our heads on them, we also had all the advantages of our labour-saving gadgets that could now be used. Madge had an electric vacuum cleaner which did a quicker and cleaner job than dustpans and brooms. It was also a source of great delight to Charles who, when one was cleaning upstairs, would sneak along to the main plug downstairs and switch the power off.

We didn't have a washing machine but there was an electric boiler in the engine shed, which was a considerable improvement on soaking and pummelling sheets and filthies. An electric iron completed the job. The boiler, incidentally, was a great improvement for bathing, which was done these days in the engine shed, for here was this huge quantity of hot water right beside you to luxuriate in. No making do with a couple of buckets carried from the house.

All that was needed now to bring us entirely up to date was water in the house. This did not come for another 18 months, at round about Christmas 1952. Even then it was not mains water; only our own sweet well water piped in. It was another 10 years before mains water was laid on round the district, and a little less for mains electricity.

Madge had a story she liked to tell of the time during the war when she was Welfare Officer to a big munitions factory in Bridgend, South Wales. One of her jobs was to rehouse bombed out and other displaced people. Many of these were evacuees. Some had been drafted into war-work from other parts of the country, as indeed Madge and Cecil had been. Accommodation was primitive and scarce for everyone, local people as well as incomers, and part of Madge's job was to sort out priorities of hardship. One of the questions she had to ask was, 'Have you got running water?'

A great many hadn't. They'd been placed in semi-derelict country cottages or slums, anything with a semblance of four walls and a roof. Much of the housing, even in bigger towns like Bridgend, was still without piped water. The circumstances of these people's lives were all very similar, but were told in many different voices of anger or desperation or dramatic Welsh rhetoric. It was all very harrowing and depressing.

Then in the middle of it all, she heard the sound of a voice from the East End of London answering her:

'Running water, Lidy?' it said. 'Yes we've got running water – it's running all over the bleed'n place.'

Madge sat back and enjoyed a rare roar of laughter.

She never thought then that she'd soon be spending a whole six years of her own life deprived of piped water and its associated comforts. Not only that, but just as the

Cecil offering a drink from the surface well, to which the spring from the sitting room runs.

Cockney gentleman had described his house, water here was running all over the bleed'n place. It hung in the air, it ran down the hill around us, it crept up through the walls and down through the roof. Inside the house it grew black moulds and caused the paper to fall off the walls.

The atmosphere was, by the time of piping in the water, rather better than it had been, not least because the sitting room was now lived in every day and had a fire burning all through the winter. Madge had the room plastered, too, around 1949/50, and underneath the plaster was waterproof cement. I think there was still an occasional damp patch, but the room was now much more cosy and bright and almost suburban!

One little quirk remained for which there was no easy remedy. The settee was on the left-hand side of the fire and so the corner seat nearest the fire was always well sat in. I had noticed that it was getting rather worn and saggy. No amount of plumping-up of cushions seemed to help this slow list to port. And then one day, almost unnoticed, the whole end collapsed through the floorboards. They had rotted away. Underneath was a small stream. It rose from a spring in the corner of the room, ran diagonally across the room, under the outside wall and on under the yard till it came up again in the well opposite the front door. There was nothing to be done about it except keep renewing the floorboards every few years. If some idiot doesn't fill it in, it may come in handy again in some time of future disaster. People barricaded inside against marauders would have unlimited fresh water.

I missed most of the work and the excitement of the water works because I had decided I needed to get away and see a bit of life elsewhere. I chose the winter because normally that's the quiet time on a farm. Also that year Flicit decided to give up permanently the various research jobs she earned her living at and move to the farm to live. So she was there to help.

I had signed up for a course in shorthand and typing in Ipswich. Then for three months after Christmas I intended going down Devon way for some sort of a job and a look round this unknown countryside.

On the way to Ipswich, near Burford on the A40, my scooter broke down – surprise surprise! – and two very nice young men on a motorbike stopped, produced a rope, and towed me to the nearest railway station. Incredible! I said could I give them something for their time/petrol and they said, just get a rope and do the same for someone else sometime. I did not literally do that. I don't think I could manage it; but I have done equivalent things many times since.

While I was away, all this water business started. I had thought it was mostly finished and baths being taken by Christmas 1952, but looking through some old letters I find it was not until the beginning of February 1953 that all essentials were completed and the bath itself ready to be used.

The preliminaries went well enough. The laying of the water pipes was easy because the way from the well to the back of the house was through the soil of the kitchen garden. No base rock to be hacked through. All sorts of people came from round about providing casual labour, and if it had been only a matter of piping in cold water to the kitchen and a water-closet, that would have been that and soon done.

But since we had already got electricity, we were obviously halfway to having running hot water as well. And with running hot water, we could have baths. So a bathroom had to be built. There had to be more pipes and taps, all connected up to a bath and a hot water tank. Madge bought a smart little stove to have in the living room in place of the old open range and this was also connected to the hot water system, supplementing the electric heating. A lot of heart-searching and discussion went on about spoiling the authentic look and feel of the old room by having a modern stove there, but it turned out to be so efficient and cosy that we all quickly agreed it didn't look the least bit out of place.

To contain all this hot water, a copper hot water tank was put in the corner and shelves and a cupboard built around it, making a fine airing cupboard. There was a thermostatically controlled immersion heater and the tank was always referred to in the family as the immersion tank.

When I went home for Christmas, the only additional convenience was cold water on tap to the kitchen sink. The shell of the bathroom had been built beyond the kitchen, sadly destroying a fine victoria plum tree and a beautiful old weigela to make room for it.

I went off again immediately after New Year and found myself a job as a pantry maid in a hotel just outside Exeter. This seemed ideal as it provided food and accommodation and bits of free days here and there when I could go exploring. It also gave me an interesting and intimate experience of all the strange goings-on and the different layers of privilege there are in 'downstairs' society. In addition to all this fun, the hotel, tucked away in the countryside, was the place where the Assize Court judges and all their entourage were accommodated during their Circuit of courts, when they were trying cases in that part of the country. And they were in session within the first week or two of my arrival. I was too low down the staff ladder to do more than catch glimpses of these great men and their wives, though the lesser beings, their counsel, the circuit barristers, did fraternise a bit and even gave me a lift one day when I was walking back

Festive sitting room with electric light. The well across the yard is slightly to the right. The underfloor spring runs diagonally from the opposite corner.

from taking my poor little Vespa for yet another new clutch. The closest contact we had was in the back kitchens where we entertained lower members of the judges' staff, including their own cook and a couple of policemen. This custom of travelling Assize Courts stopped a few years later, so I was particularly lucky with that adventure.

While I was enjoying all this luxurious entertainment, there were such great problems and disasters going on back at the farm, I'm surprised I didn't give up and go home there and then. Cecil was quite ill, Flicit had terrible toothache which couldn't be cleared up quickly, and young men of the village who would normally have helped out were having to stay at home because their wives were involved in difficult childbirths. I gather from Madge's letter that I did suggest coming home, for she says I'm not to.

It was in the first week of February 1953 that the new stove was lit and hot water piped to the kitchen sink and the bathroom. I bet there was a boozy celebration of the turning on of hot taps. And even more so for the first baths on Sunday 8 February, even though there was no bathroom door. There remained no bathroom door for two or three weeks. Madge was always afraid our neighbour from up the road would wander in while she was in there, asking to use the phone.

However, as I said, this was not mains water and though it now flowed freely in the house at the magical turn of a tap, the cold water still had to be pumped by hand from the well to the standing tank. It was from this tank that water flowed when the cold taps were opened and it also refilled the immersion tank as the hot water was run out. This was the standing tank that also supplied water to the cows when they were shut in for the winter. If the tank was allowed to run dry, thirsty cows would soon be bellowing and bashing furiously at pipes and water bowls. They were not kept waiting long.

However, failure to pump water to refill the immersion heater at bath time would bring an even noisier response – an explosion. So no one ever forgot to pump – though remembrance sometimes came belatedly in a frantic screech from the bath!

So how was it that one day a few weeks later, hot water went out of the immersion tank and cold water failed to come in?

After the almighty bang that shook the whole household, water was well and truly all over the bleed'n place. The door was blown off the cupboard and all the clean clothes airing there were doused with water mixed with cobwebs and loose plaster off the walls. Water gushed out over the living room floor and into the kitchen. It ran and ran. A lot of very rude things were said in loud and angry voices as we all rushed round with cloths and buckets, mopping up and building barricades.

It was quite a little while before someone suddenly realised there was an amazing amount of running water coming out of what should have been an empty tank. In fact, far from being empty, the cold water tank was full and doing a great job refilling the now non-existent immersion tank.

After a lot of frantic searching, a faulty safety system was found. This was remedied and a new hot water tank and heater were installed. And we all bathed happily ever after.

The remains of the copper tank were tucked away among the Things That Might be Useful For Something Sometime. Later they passed to my husband, Patrick, who used bits for this and that, most notably a smoke canopy over the fire in our present home.

Chapter 18
Cliff Hangers

IT was one of those mornings when I rushed over late from the tent with only an overall on and had to go immediately to take the milk up the lane. Some kind of crisis was going on which made everything late and awkward. Perhaps a cow was having difficulty calving. I don't remember. What I do remember is taking the churns as David lifted them onto the back of the cart and rolling them one by one to the front. I had pulled my overall modestly across my breasts and stomach, tying it tightly together with a piece of binder-twine to be sure of not shocking David. Only as I was urging Bess away up the yard did I suddenly become aware of a coldness around my bottom reminding me of the slit up the back of the overall. I sat hastily on an icy-cold wet churn, glancing back to see what David was doing. It was no great consolation to see no sign of him. I stood up and tried to adjust the overall more evenly around me. There was still the likelihood of meeting the driver of the milk lorry ahead, but with luck young Joe Hakesley from the farm up the road would have been and gone by the time I got there.

There were no new morning churns on the stand, so that meant Joe was also late and would be rushing down any minute. I drove Bess across the road then backed her so that the back of the cart was against the edge of the milkstand.

'Whoa Bess!' I said, trying to put on David's authoritative horse voice. 'Stand girl.' And I pulled the reins over the front of the cart and tied them, as I had seen David do, which was supposed to make her think I was still holding them and in charge.

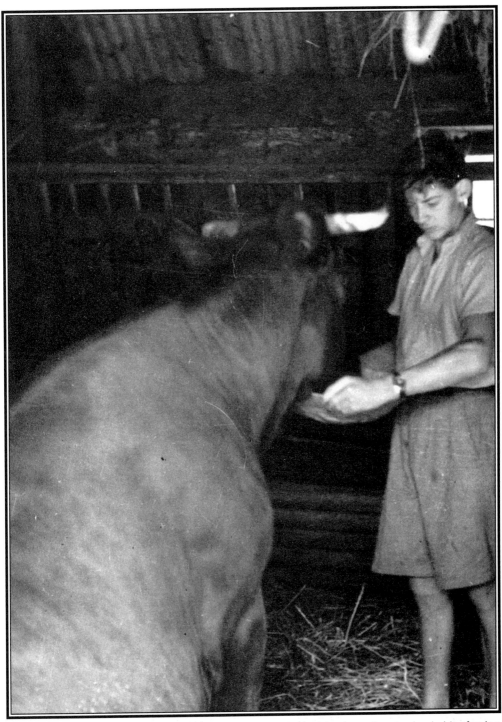

Cow who thinks she's a horse. Ken offers food to Cochen, stuck on her haunches due her habit of trying to get up front legs first.

Mucking out the stable. The ergonomics are all wrong: the barrow should be turned downhill, the way I'm going, before filling.

Milk churns were being crashed about along the road, indicating that the lorry would be along any minute; and at the top of the hill straight ahead, another horse and cart came in sight trotting my way. If I could hurry and get the full churns off the cart and the empties on, then I could remain sedately seated when the others arrived. And why should it even enter anybody's mind to wonder if I had anything on under my overall?

I moved the three churns to the back of the cart, put one foot on the milkstand and easily swung the first churn across; then the second; and the third. No sign of the lorry yet and Joe was hidden behind a bend in the road. I got onto the milkstand and rolled my full churns to the back, grabbed two empties and stretched across to put them onto the cart. As I swung the first one over, I heard the lorry coming along the road and at the same time I felt the cart moving away taking my left foot with it. The right foot was still on the milkstand. I yelled at Bess to whoa! and began to perform a slow splits. At the same time as my legs stretched apart, my body was swinging forward into a deep bend with the weight of the churn I was holding. I heard the lorry's brakes but luckily my back was not turned in his direction. A quick decision was needed. I dropped the churn and managed to push myself into the air while my feet were still close enough together to put pressure on them. I landed successfully, reasonably sure nobody had seen my undignified position. The milk lorry was waiting and Joe had run to the horse's head to hold her. I slung the milk churn on the cart, climbed up for the other two and took the reins.

'Shall I come and hold Nell?'

'No. She'll be all right.'

As I pulled Bess and the cart round to drive back down the lane, I found there was a car waiting to pass. I couldn't distinguish the face inside, or its expression. Oh well, what the hell. No point speculating. I waved and smiled cheerily. Joe was backing his cart up to the milkstand and as I moved on he put two big stones against the cartwheels. I had put stones under the wheels the first two or three times I'd gone up, but then I saw that Bess never moved when David ordered her to stand, or when she occasionally did move there was no harm done, so I stopped bothering. I did have one or two other awkward moments, but none with so many elements of farce as on this trip.

There were usually spare fine days between hay and corn harvest when we might drive out to visit places of interest farther away. But mostly we went in fours or fives, eights or elevens, with picnics and bathing things to Penbryn beach. We almost always walked the three miles each way. Only Cecil would come later in the car, and sometimes Madge, joining us for a picnic and maybe a couple of hours of loafing about. One or two people might go home in the car, either from tiredness or because there were jobs to be done, like animals to feed and milk.

They were lovely, enjoyable days, though mostly we only swam a bit, sunbathed and talked. We walked the length of the beach and round the rocks to the second bay when the tide was out, and we played quoits and cricket. In the second summer Madge bought a rubber dinghy left over from the war, and we had tremendous fun with that. Mostly it was me and Madge and Hilda, and we wallowed in the waves, splashing each other and trying to tip each other out. We sang, 'Pull for the shore, sailor, pull for the shore.' Hilda would grab at her ill-fitting bathing costume from time to time with a cry of, 'My dumplings are boiling over!' Doesn't it all sound so simple and childish! There were no transistors on the beach or speedboats in the water, but I had many fascinating talks about music, poetry and philosophy on those long treks to Penbryn and back.

On one occasion, a girl about my age wanted to climb across the upper cliff above the two bays. It didn't seem very interesting to me. It was high, but involved only a struggle up through scrubland and bushes. The way over the top looked a straightforward grassy slope, but knowing these cliffs as I did, I expected it to be more complicated than that. I had the feeling she thought me too puny for the task and it was a bit of a challenge between us.

The terrain turned out to be even worse than I'd anticipated. Not only were there thickets of low growing blackthorn to get round and through, but between them the ground was treacherous with steep scree slopes and loose tussocks of grass. They needed to be crossed or skirted round with great care, but this girl went stomping along as if she was on flat and solid ground. She was a great walker and used to tramping up highland tracks, but she had no feel for the delicate structure of slate cliffs with thin soil and shallow rooted plants. I don't remember if she had any slips. I do remember being terrified that she'd go sliding and bouncing down to the bottom at any minute. We were probably between 100 and 200 feet up and there were rocks at the bottom.

And her mother!

I was assessing the way ahead as best I could. My friend seemed to have no conception of any possible difficulty and was determined that we were going to get to the top. I wasn't so sure. I hadn't climbed these particular cliffs but I had done a lot of climbing around Llangrannog and I knew that the top edges were generally difficult and dangerous. There was not a lot of choice about the way we were going. Impassable bushes or unclimbable gullies directed us towards only one possible way forward. I looked back at two tiny figures on the beach below. We were now at about 300 feet. The thought of going back the way we had come did not appeal. It had been a long, hard climb and would be little better going back. The top was within reach, then a gentle walk back across fields.

Confronting us before the top was an almost vertical wall of soil with slate shards sticking out of it. Odd roots and clumps of grass and heather were growing here and there. The top was above our heads. My companion grasped a couple of clumps of grass and stuck a toe on a chunk of slate. She was poised in this position for possibly two seconds before the whole lot crumbled. I steadied her as she fell back, worriedly glancing down to the beach and thinking the two figures looked agitated, though I'm not sure on what I based this conclusion.

'Errmm...' I probably said – or something like it.

My companion moved along the cliff a bit and tried again, a little more carefully. She was no more successful and for the first time in all of this expedition, she turned to me and asked what we should do. I had already worked out where I could almost certainly get myself up, but I knew she could not.

'I'll put my hands for you to step up on,' I said.

'I can't do that.'

'Yes you can. That'll put you high enough to reach over and pull yourself up.'

'But what about you?'

'I'll be all right.'

I held one arm against the cliff and supported it with the other. She stood on the back of my clenched hand.

'Reach as far as you can. The roots'll be firmer away from the edge. Grab the strongest plants and try not to put all your weight on them.'

As she leaned forward I pushed on her feet and there she was, over the top. She stood up and came forward.

'Don't stand on the edge!'

She stepped back.

'I'll give you a hand.'

I told her she'd break the edge of the cliff away and that would weaken my hand and foot holds. I didn't think I'd have any trouble. I was lighter than she was but I also knew which were the strongest holds and how to distribute my weight to avoid putting too much pressure on one place. And so we returned safely.

I caused even more consternation on the cliffs a few years later. I had a special place in Llangrannog I liked to go sometimes. It was entirely secluded, except from the sea. I had to walk right along to the Ynys to find anywhere comparable, and anyone else a little bit energetic could do the same. Not, though, with the place Madge and I called the Mermaid Rocks. This is to the left of the village, round the other side of a grassy,

Left side of Llangrannog Bay, Pen Rhyp, where I caused consternation by disappearing over the edge.

pyramid-shaped piece of ground standing out from the bottom of the cliff. A small inlet and mounds of rocks made a splendidly sheltered, private place to dream, swim, read and sunbathe. Waves crashed and splashed; birds called in windy voices; an occasional seal's head would pop up and ride on the swell, very still and seemingly disembodied; thoughtful. Now and again something black and white would flash down the cliff face and disappear into the sea. I couldn't understand it the first time it happened. It seemed to be a bird, but birds don't dive into the sea. So I thought it had fallen or died. I can't remember now whether it would have been a razorbill or a guillemot. Perhaps either? I do know both dive like ducks from the sea's surface.

I swam round to this place first of all because I needed to explore the climbing possibilities from the bottom before I could try coming down from the top. That way I would see possible tracks up, and if I got it wrong, I could go back down the way I'd come and try again. As far as I remember, I made it first try. It was straightforward and not too difficult. Until I reached the bit before the top. Here was the high straight face that is the shape of most cliff tops immediately under the edge, often with an undercut.

Behind me was an almost vertical drop of around 300 feet. True it was grassy, but a body would be unlikely to stop before it had bounced onto the rocks. I felt my way along, testing the tenacity of plants and the hardness of patches of ground that offered foot room. They are a strange mixture, these cliffs, parts packed hard, almost like cement, then loose soil where water has seeped through and bits of slate have moved. I expect there is some more sensible, scientific description, but whatever, these cliffs are treacherous. It took me a long, worrying time to work my way up, and when I sat safe on the top, I was quite sure I was not going to be climbing down. But I did, many times.

One day I walked up the road to the cliff top and a farmer was working his way backwards and forwards across the fields with his tractor. Not wanting to worry him by letting him see me climb over the sheer cliff edge, I waited till he was out of sight. After a couple of hours I climbed back and called in at the Pentre Arms for half a pint before making the long plod up the two hills to home.

'Have you been up over Pen Rhyp?' said Gareth.

'Yes.'

'There's half the parish out looking for you.'

'What for?'

'Idris said a small boy in short dark blue trousers had gone over the cliff while he was up there harrowing.'

'Oh Lord!' I wailed. 'I waited till he was out of sight in case he thought that.'

'That was the trouble. He said there was no time for the child to go anywhere but over the edge from the time he saw him till he came back and the child was gone. I told him I'd seen you go up there about that time and he needn't bother about you, but he said no one could climb down there. He's got all the farmers between here and Penbryn out looking for you.'

I groaned and set about letting them all know I was all right. I walked round the farms next day, six miles or more I suppose, to apologise and offer to do some work for them to help make up their lost time. They were all very nice and saw the funny side except for one. He moaned at me about wasting his time and effort, called me irresponsible and that sort of thing, refused my offer of help in tones suggesting I'd be useless anyway. I did answer him back mildly eventually. He was the only Englishman among them!

Chapter 19
Corn Harvest

THE word 'corn', in England, used to be a general name for wheat, oats and barley. What is now called sweetcorn was known as maize before the war and was used in chicken feed.

In Britain, corn is sown in spring or winter. Winter corn was rarely grown in West Wales. Usually the land was too wet for cultivation in October and November. Wet land would clog in the plough and not break down to a proper growing tilth. The grain might even rot in the ground from the wet. The great advantage of winter corn, when you can grow it, is that harvesting is done in July at the height of summer. Sun is hot and days are long. Ideal for big, corn-growing farms. In our area, long dry spells were rare.

Apart from climate problems, our small farms, on average 40 or 50 acres, could not afford to lose land to winter corn for about 18 months. This is the time taken for the corn's growing and harvesting and the field's resowing. If corn was sown in October, cleared in July, the field ploughed and sown with grass again in late August to September, it could not be grazed until the following spring; possibly earlier with sheep. With spring corn, on the other hand, the field wouldn't be ploughed and sown until March, which leaves grass for grazing through October, November, and even later if the weather is mild. The corn can be undersown with grass, which provides lush grazing for use again by October – with any luck! – immediately after the corn is taken in. Winter

The binder cuts corn and ties it in sheaves. Something has gone wrong here as sheaves are coming out very messily. The wheel and chain mechanism can clearly be seen turning the paddle wheel.

corn cannot be undersown because the grass would grow as tall as the corn. Even low growing spring grass could be a great nuisance because it packed thick in the butts of the sheaves and took a long time to dry. You could have fine weather enough to dry and ripen the corn, but couldn't cart and stack it because the fresh grass in the butts would make the rest go mouldy. In the end there had to be a compromise and when stacks were being built, the greenest of the sheaves would be shuffled around to be stacked into the outside walls, their grassy butts safely out in the air. Others would go on top in the steep watertight capping.

We started to worry about harvest in July. If the weather had been fine and haymaking was finished by the early days of the month, Cecil could take his time getting ready for the next lot. There would be days out to castles and mountains, rivers and churches in the north and south of the country. And there were those days walking the cliffs and beaches. But niggling away always was the worry of the corn and the weather to come.

Towards the end of the month, the great cumbersome binder was brought out from its shed. We were lucky to have our own binder. Some farms depended on borrowing from neighbours or hiring from the War-Ag. Either way you were greatly disadvantaged. Somebody had to be, because when the crops and the weather were ripe, everybody needed the machines at the same time and there were not enough to go round.

Galloping (Fordson) Major. Better sheaves here. The 'tent' covering mechanism is seen with prongs that pack and push out the sheaves. The crop is very tall oats, awkward to cut.

The first thing to be done was to check all working parts for damage, especially the delicate blades of the reel: the big open 'paddle-wheel' that guided the corn onto the bed of the binder. Nipples and chains were greased. Two knives, already sharpened and greased over winter, were loaded, plus the two canvases of the conveyor belt. The machine, when it was set up for cutting, was too wide to move through gates and along the lane, so it had a cunning system for travelling sideways. The tow bar could be taken off and fixed at either the front for working or the side for moving around. Once in the field, the tow bar was fixed in the working position, the towing wheels were removed, and the knife was set in place at the front of the bed. The two canvases of the conveyor system were strapped tightly in place. One of them lay almost flat across the bed, the second rose at nearly a right-angle to it, to carry the corn to the top of the 'tent' which covered the machinery. Each of the canvases moved around two rollers and had slats across to support the corn and stop it from slipping back. Two balls of twine were set one above the other, in a cylindrical container at the back end of the 'tent'. An end of twine was extracted from the centre of the bottom ball, just long enough to be tied securely to the outside end of the top one. Then the inner end of the top one was fed out and through various holes, including the eye of the sheaf-tying needle. As the sheaves were tied, the twine was slowly used up to its last little end and so to the knot that brought the second ball of twine into play.

Before cutting, Cecil would be making trips through the corn every few days to check how it was ripening. And there is another lost skill. How many farmers nowadays, even including organic farmers, would know the rules for harvesting corn with a binder or scythe should the need arise? How many could walk through a crop and judge by the colour of the straw and grain, and the feel of the grain, whether or not it was ready to cut in this way? For corn to be gathered in by the flailing sails of a binder, thrown onto the conveyor sheet, sorted and packed into sheaves which are then thrown back on the field, it must have sturdy flexible straw and secure heads of grain. Quite different from corn gathered by a combine.

And how many would know how long then the sheaves had to stand out to dry? They should hear the church bells ring three times, was the old country saying. Barley ripens first. Stalks crack and the heads of long-whiskered seeds nick over. Oats are next, the husks beginning to gape as the grains ripen. Stalks had to be slightly green and ripeness was tested by feeling for plenty of nice plump grains and finding a milky fluid inside when they were cracked open with finger and thumb. This fluid hardens into flour when the grain is fully ripe. It was customary in West Wales to grow some oats and barley together as a combined crop. The longer stalked oats supported the much shorter barley, keeping its heads off the ground in the least of the bad weather. Barley ripened first, late July to early August, oats were a week or so later. A compromise had to be made over cutting the mixed crop to allow for the slight differences in ripening times.

Wheat is cut last, towards the end of August. It is the most beautiful looking of the crops, standing tall and strong with shades of green and red in the stalks and a rich gold in the heads. The straw, though, is useful only for thatching. Wheat was grown here only by government order. It did not produce a quality grain, soil and climate being too poor, but the War-Ag never managed to deduce from this that it was pointless to insist on these farmers growing the regulation two acres intended for national consumption.

The binder was an incredibly clever machine. As the stalks of corn were cut, they were guided by the reel to fall onto the canvas conveyor belt with their heads all lying the same way. They were carried across the bed by the first canvas then up the steep slope of the second. Here they continued in the same tidy layer, over the edge and down through a slot into the inner workings. They slid on downwards assisted by a couple of revolving prongs and were packed against two sloping boards which sat end to end with a narrow gap between them. At the same time another device was banging on the straw ends of the bundle to make a sloping 'foot', essential for getting a firm stook and for building a stack. When the bundle was the right size to make a sheaf, the packer prongs

Field of stooked oats. Looking over the village, across Cardigan Bay.

stopped pushing and a needle and twine came up through the gap between the boards, wound around the sheaf and tied the twine in a knot. Something cut the twine, an opening appeared briefly for the sheaf to shoot out through, assisted by the prongs, which were now turning again and already pushing the next stalks down for packing and tying. Each whole operation took only a few seconds.

Somebody sat on the binder to check that all was going well, and also to raise and lower the cutting bed and the reel when there was a change in the lay or height of the corn. Madge liked this job, which I think she and Cecil always did together, Cecil driving the tractor. There would always be two or three problems during the cutting. A blade might break on the reel, or the twine might snap. A considerable quantity of untied corn would land on the field and a considerable number of loud, rude words would pass into the atmosphere before Madge's frantic shouting got through the engine noise and Cecil stopped the tractor.

You didn't get this problem with horses, of course, but we never used the binder with horses. Big machines like a binder needed at least two horses to pull them, probably three. The two were harnessed one each side of the pulling pole. This pole was held up at the front by a short crossbar attached to the bottom of the horses' collars. Behind the horses, a crossbar was loosely fixed to the pulling bar, allowing it to swivel slightly so

Ken with Henry building sheaves into field mows.

that one horse could move farther forward than the other when they turned. They were attached to this by their traces and whipple-trees.

Very little skill or knowledge is required to judge when corn and weather are right for combining. The corn must be quite ripe and dry, the day fine, and then the whole year's crop is not only cut, but threshed and put in store within the day. Combines can save grain that is so ripe it sheds at a touch. And they can pick it up when storms have battered the crop flat on the ground. The old binder could not do any of this.

There's really very little fun in combining at all! Two days at most of hectic rushing around and it's all over. No long, leisurely days walking up and down the fields collecting sheaves and standing them up in neat rows of straw houses under bright blue, sunny skies, looking out over the western sea to far mountains.

In fact, come to think of it and taking off the rose-tinted specs, there wasn't much of that either. We were invariably hurrying because there were several fields waiting. If it was fine, you never knew how long that weather would last; or if it wasn't so fine, then you were hurrying to at least get the sheaves stood upright to shed the rain. Later there would be the rush to make mows, small stacks built in the fields and containing fifty or sixty sheaves each. These protected most of the corn from all but the heaviest and most persistent rain. There were lots of fine days though, and even if we had to work hard and fast, concentrating more on the work than the idyll, it was still mostly enjoyable work.

And again, there was a right way to stand the sheaves up. Stooks in our area were mostly made with four sheaves, but sometimes six were set in a row. You picked up one in each hand and stood them head to head with the short slope of the 'foot' inside, leaning them towards each other. You thumped them down fairly firmly and you pushed the heads together firmly. Then you got two more and stood them at a similar angle each side of the first two. The stook had to be stable yet stand as upright as possible to shed rain and to avoid subsiding into a heap. There had to be space also for wind to blow through and dry it. I was taught that if the first two sheaves didn't stand steady on their own, neither would the completed stook, whether with four or six.

And all of these jobs had to be done not just the once but two, three or even more times. Winds frequently blew stooks down during the weeks they were standing out. Since they invariably had some rain on them too, stooks and mows had to be thrown open when drying days came, then if they didn't dry enough or there wasn't time to cart everything home, they had to be all put together again. The sheaves were tested for wetness by thrusting a hand into them. The tightly bound centres were the last to dry because weeds kept fresher and greener there. As a lot of the weeds were thistles, many of the sheaves were also prickly. Hands and fingernails became very tender with thistle thorns.

The autumn equinox and its gales always hit us at some time during harvest so there were many times when the gales swept flat all our neat rows of stooks.

But there were other hazards too.

Cows!

As with kale, only more so, there were few things in a cow's life so enjoyable as getting into a field of stooked corn – and few things in a farmer's life so infuriating, frustrating and exhausting as trying to get them out. For of course, it is never just one cow. Cows are gregarious, sharing, communicative people. When one finds a hole in a hedge, she does not just sneak quietly through and get on with it. She shouts out, 'Look what I've found, girls!' and they look, and they kick up their heels, and they charge after her, pushing and shoving to get in first. Where there had been a couple of stones loose in the bank and a small gap in the hedge caused by a branch being broken, there is very soon an earth track leading through an opening in the bushes as if a small tank had passed that way.

And if they found themselves in a field of stooked corn, then it was carnival time. They might grab a mouthful from the nearest sheaf, but then it was heads down, udders swinging, and up and down the rows they went, chewing, tossing, kicking. It wasn't

Cecil gathering loose corn to tie into sheaves.

because you were trying to chase them out. They just ran for the hell of it. The game was to demolish as many stooks as possible, and it was hilarious. They were so funny. And so happy! There is really no other word to describe this mad hullabaloo. Perhaps there is some happy drug given off by drying corn and its associate herbs. Even we humans had to laugh at least once through our fury and breathlessness to see these normally stately, rather cumbersome animals galloping about, kicking their heavy hindquarters in the air. Their great milk bags were shaken about so much you might expect to find butter in them at milking time. Once in a while one would stop and pose as a music hall duchess with a floppy straw hat skewed over one horn and the remains of another sheaf drooping from her mouth like a tatty fag.

Cows don't seem to break out of their fields nearly so much these days as they used to. This is largely a matter of shortages again: time, money, materials. Fencing wire was scarce for some years after the war and we were for ever boggling bits of rusty barbed wire together to try to block a hole. I had many nasty cuts from the stuff. Fields were divided by high banks made of stones and soil and grass. Trees and bushes grew on the tops. Cows naturally chewed the grass, climbing the banks to reach the higher bits. They also rubbed their heads in any earth that appeared and dug their horns in. Without even intending to, they soon found themselves in the field next door. A strand or two of barbed wire was fixed to posts and trees along the tops of banks to stop the climbers, but this got broken or was undermined. In much later years, Patrick just put barbed wire fences round all the fields to keep cows and banks apart.

There was one year when harvest conditions were absolutely perfect. No gales, no rain and no cows. The entire crop was home and stacked before the end of September. Amazing! And a few days later, we had the first threshing. The straw was lovely: golden dry. And the grain was dry and heavy.

But nothing, of course, is ever utterly perfect, and halfway through the gathering, while Madge was driving one of the tractors, her foot slipped off the clutch and gouged a great lump out of her leg. It was very nasty and people wanted her to go to the doctor to get it seen to.

'Not bloody likely!' said Madge, knowing that would mean stitches and stitches would mean keeping the leg at rest. She went on driving the tractor or riding the binder for several more days with miles of bandage round the leg. We probably would not have beaten the weather if she hadn't done so, for it needed all hands going full blast. We were lucky that Hilda was staying with us, helping with everything as always. She had been due to go home but got in touch with husband Joe to beg a few more days. All was almost gathered in, and then rain came. We had just nine or ten mows left out in the field, five hundred or so sheaves.

It wasn't a lot of rain and halfway through the day the sun came back. Tractors and trailers rushed out again and the last sheaves were brought in. The peak of the stack was shaped and ready to be finished off when filthy black clouds appeared and a violent wind blew up. Our stacks at that time were not thatched to make them weather proof. We cheated by putting tarpaulins over the outside ones. A great battle took place to get this last one covered against the gale and imminent rain. Our butcher arrived as we were working and he stopped to give a hand, as did another man who had called about something. These two went up on the very precarious stack. It had gone up very steep like a church steeple and they seemed to be hanging on by no more than a few strands of straw. They somehow managed to work the sheet over the top. As they pulled it round, they threw down the ends of the attached ropes for the rest of us down below to catch and hang onto. The way the whole thing tugged and danced about reminded me of a tethered barrage-balloon in the war, tugging and bouncing to get off the ground. As we hung there wondering what to do next, Madge had the brilliant idea of getting the spare gates that stood in the yard and tying the ropes to them. Which we did. They were heavy oak gates and they kept the tarpaulin safely in place through the equinoctial gales which blew up next day and roared around the place for nearly a week.

Madge's leg took a long time to heal and left a significant scar. She used to show it off with great pride.

Chapter 20
Cardi Connections

THERE are no precise outlines of West Wales and Mid-Wales, as far as I can make out. They seem to be as variable as the weather forecasts whose borders move in and out to accommodate the spread of rain or sunshine. However, Ceredigion fits somewhat into both. It is also definitely South, both from a cut across the middle and from a cultural point of view. Rude remarks may be heard in North and South Wales regarding the status and abilities of people on the other side of the divide, as happens in so many countries.

Ceredigion, or Cardigan, has always been one of the poorest areas of Britain. The soil is mostly shallow and rocky with large areas of bog-land, and the climate is fairly consistently wet. Commercial centres have always been far away in terms of time and convenience of access, and despite all technological advances, this still remains true in many respects. The fact that the word 'Cardi' is still a byword for meanness throughout Wales is a sure sign of long-term poverty and consequent carefulness with money. But mean in character and neighbourliness they certainly are not. We found them always ready to help anyone in trouble and I knew of two cases near to us over the years when farmers were too ill or badly injured to work and neighbours put together a rota between themselves to keep the farms working. One man many years later had his arm torn off by a revolving shaft on a tractor and neighbours helped to keep his farm running for over a year.

I had a lovely personal experience of Cardi kindness on one of my meanderings. I was coming home from Nottingham in January 1948 and I forgot my instructions about which station to go from when I changed at Birmingham. I could go from one – I still don't remember which was which! – to get to Carmarthen, which I wanted. Or I could go from the other – I rather think that was Snow Hill – to get to Aberystwyth which I didn't want.

Well, it was a very slow train to Aberystwyth, and when I got there the last bus going Llangrannog way had left some time ago. This was seven or eight o' clock-ish on a Saturday night and there were no more buses until Monday morning. I had no money. Well, a few bob for bus fare and a packet of crisps. I tried to explain this to the station master but he wouldn't listen. He was not having me sleeping in the waiting room. He insisted on walking me over the road to the Commercial Hotel where, after a few words in Welsh between him and the proprietor, he left me. I tried to explain again about my mistake and my lack of money, but the man said it was all right and I could post the money to him when I got home. Which of course I did. One of my happy memories of life and people.

That Sunday spent in Aberystwyth would not go on the list of best days though. The beach was cold and grey and wind-swept and the only people I saw were stiff and black and either going in or coming out of chapel. At least nowadays the pubs are open on Sundays, possibly even an occasional shop and cafe. Cardiganshire was one of the two last Welsh counties to stand out against Sunday opening for pubs. For several years there was a sign on the Machynlleth to Aberystwyth road, just inside the Montgomeryshire border, which said, 'Your last chance for a drink for 60 miles'. Sixty miles would have taken you just over Cardigan bridge into Pembrokeshire. A referendum had been held every five years since the early 1980s, in which people were asked to vote for whether their county should be 'wet' or 'dry'. Each time more and more counties voted for Sunday opening until only two remained adamantly dry: Cardiganshire and Lleyn Peninsular. When the next referendum fell due, the government said it wasn't spending more money on yet another referendum just for those two counties and they would have to conform to majority opinion. But that great Sunday revolution started nearly 40 years after my dreary but lucky Aberystwyth Sunday.

There are also now Sunday bus services, though sadly most of the lovely old country railways have been closed down and dismantled. Bus services in those days were frequent and well used. That between the railway station in Carmarthen and the beach at Llangrannog was not only convenient and efficient, it was adaptable to all sorts of hazards and delays. If a connecting train was late getting in, the bus would wait. I know of one occasion when it waited nearly half an hour, but that was rare. So connecting buses at Newcastle Emlyn, halfway between Carmarthen and Llangrannog, also had to wait. There were some rather heated disputes sometimes between the two drivers and some of the passengers, fed-up with being kept waiting about. Dependability of the connections was more important than a little irregularity at the other end of the journey.

Trains would wait for buses as well, though not for more than a few minutes. Madge and I were very glad of this on one fraught occasion. We were setting out for a holiday in Austria and had a whole series of connections to make. Then only three or four miles from home a part of the bus fell off! It didn't break off altogether but bounced and dragged along the road. The conductor and the driver looked at it and tried to put it back and a lot of chat went on. Madge was getting somewhat agitated by now and had various words with them about the urgency of our situation and couldn't they possibly tie it on or something. I don't know how long we stopped there, maybe 10 minutes or so. Then the conductor stood on the step by the door and the driver passed him the bit of mudguard. And there he stood, hanging on to the mudguard with one hand and some more stable bit of the bus with the other, while we bounced and bumped and swerved along these rough roads for another 10 miles or so. All connections were made!

One day Felicité was setting out from the farm to catch the train home after a holiday. Cases were on the doorstep and she and Madge and I were about to start off up the lane to meet the evening bus. I think it was due to leave the village at six o' clock and would arrive at the top of our lane at about ten past, after driving round via Pontgarreg, picking up various requests along the way. It took us roughly the same amount of time to amble up the lane carrying the luggage and we were setting out with five minutes to spare. The bus would not stop if we weren't there to wave at it.

Then Flicit made a disastrous discovery. She had no cigarettes! At a time when very nearly everybody smoked nearly all the time, the prospect of a couple of hours without cigarettes really was a disaster. Madge had only one or two and needed those in case she couldn't get more, Cecil was somewhere out of reach and probably wouldn't have had spares anyway, and I didn't smoke.

'I'll run to the village and get some – give them to the conductor.'

'You'll never get there in time.'

'We'll see.' And I leapt off.

It's about a mile to the beach. All downhill. Consequently our saying about the beach: It's 10 minutes there and 20 to 30 minutes back. However, doing my mountain goat act down the rough bit and running non-stop, I reached the Ship car park to find the bus still there. I shouted and waved to wait a minute, grabbed some cigarettes from Mr

Pryce-Davies in the adjacent shop and asked the conductor to give them to Flicit. She wrote later and said she and Madge had been sure I wouldn't make it in time and what with loading her things on the bus and saying goodbye, she had quite forgotten about the cigarettes. So when the conductor came along for her fare she was quite taken in for a moment when he said, 'The bus company is giving away packets of cigarettes today to all passengers getting on at Penrallt lane.' She sent me a pack of pretty hankies as a thank you, which was bountiful as they cost clothing coupons as well.

We had also a little village bus which went round all the tiny by-roads, in and out the coastal villages. I rode on it to Cardigan several times and it was a memorable journey. We rarely went precisely the same way for Wynford would wander away down farm tracks to put off elderly or otherwise encumbered persons more nearly on their doorsteps. The bus was packed to at least twice its basic capacity for most of its journey. We would stop anywhere and everywhere for people to get on and off. Wynford even waited while passengers delivered goods and messages or got off to buy an ice cream. The bus bounded along the narrow twisty roads, Wynford as often as not hanging on to the steering wheel with one hand while he turned to talk to someone or read his newspaper. I don't know of him ever having had an accident.

Cardiganshire was mainly a cattle and horse rearing area until just before the war. The sturdy Welsh Cob was very popular in towns for pulling the lighter weight carts and loads such as milk floats. They needed slightly less food and space than the great Shire horses that pulled the heavy brewers drays. They have become popular again in recent years, mainly, I think, for pony traps.

The damp warm climate of West Wales was perfect for producing grass. Beef animals were raised on this through the spring and summer. When winter started to close in and grass stopped growing, the beef animals were moved across country to the big markets and populations of the Midlands and eastern counties. These were the big corn-growing areas and farmers here bought in young cattle of a year, 18 months, two years, and put them in yards to over-winter. They used straw from their threshed corn for bedding. They fed the animals on straw and corn and then sold them on in the spring either for slaughter or further fattening. Farmers were then left with a goodly heap of processed dung and straw to dig into their fields to feed the next corn crop. This was in the years before the wholesale use of artificial fertilisers.

Another link in this chain started in a country even farther west and possibly even poorer than West Wales. This was Southern Ireland. Farmers here reared their own calves to about a year old then brought them over to Wales in the spring and drove them up through the west coast counties, selling them where they could to Welsh farmers. I suppose they were the very last cattle drovers in the country. Welsh farmers kept the stock for the few months till winter, or possibly till the following year, then sold them on for slightly more than they paid. It sounds like a good cooperative system, but it was still the richer arable farmers who set the stakes and made the profits. The Cardis and the Irish remained on subsistence living.

Two things improved prosperity and began to change attitudes in Cardiganshire and other remote farming communities. These were the setting up of the Milk Marketing Board in the mid 1930s and World War Two in 1939 to 1945. The Milk Marketing Board (MMB) was set up in 1935 to ensure that a fair market was available to all milk producers. It guaranteed to collect all milk produced from anywhere in Great Britain and to pay a standard price for all that was of basic good quality. Milk factories were set up in remote areas, churns were provided and lorries collected the milk every day from stands on any reasonably surfaced road. Farmers built their own milkstands, sometimes sharing costs and use with a neighbour. Some of the small farms sent no more than half a churn full, five gallons or 23 litres. In the factories milk was tested for levels of cream content, for freshness and for tuberculosis. All of suitable quality was either sold on in liquid form or manufactured into butter and cheese on the spot. Local communities and economies received a great boost from the business, which provided many jobs where little work was otherwise available. Any milk that failed to pass even one of the tests was returned to the farmer and he didn't get paid.

The fact that the same price was paid throughout the country meant that farms like ours, out in the far reaches of nowhere, could compete successfully with those five miles from a city centre. And so the fluctuating and not very dependable beef income was replaced or supplemented by a regular, dependable milk cheque. The Welsh Cob went out of favour, one way and another. It has recently had a revival but is now reared almost entirely on specialist farms.

One of the tests made on milk was for butterfat content. A minimum percentage level had to be reached for fat and one for solids-not-fat. Some people used to skim a cup or two of cream off their churns before sending the milk in, but we never did that. We took a gallon of milk every day, more when we had visitors, and we collected the cream off that until we had enough to make butter. We drank only the skimmed milk.

The war boosted farmers' incomes and their importance. With the need to feed the nation on home produced food, farmers were helped, encouraged, bullied, into using every fragment of land they could possibly cultivate and were told they must get two, three and four times the production from it. And so began the super-efficiency of farmers, leading later to mountains of unwanted food and unwanted farmers.

Most of the old farmers found it hard to adjust to greater prosperity when it came. They continued to live on their home-grown meals of fat bacon, swedes and potatoes, and to hoard their newly acquired spare cash in the traditional mattress or sock. Though it's unlikely that a Cardi would give up a good wearable sock and one with holes in would hardly serve. A newspaper story from as late as 1966 reported the finding of one of these hoards of money after the death of an old farmer. A cardboard box was found containing £7000 in coins and small notes.

When mains water and electricity began to be laid on around the district, very few refused it. Reluctant ones were mostly persuaded by relatives that their property would fetch less than the man next door's if they didn't have these amenities put in. Our solicitor told us a story of a distant relative of his he used to visit occasionally. Again this was quite late on in the 1960s. He called one evening to find the man in a tin bath in front of his fire. He tried to persuade him to have a proper bathroom put in.

'You can get a grant from the government that will almost pay for it.'

The man said he liked it this way.

'It will make the house worth more. People want bathrooms these days. They won't pay much for a house without one.'

'That's what Tommy said. So we got it done.'

'You've got a bathroom? So why don't you use it?'

'We did once, but time we'd got the buckets of hot water from the fire and carried them all upstairs, the water was cold.'

'Haven't you got a water heater?'

'We couldn't afford the electric as well.'

There did remain a few isolated properties where one or both of these conveniences was refused by old stalwarts. These places were later snapped up by buyers thinking they had a bargain, only to find that water and electricity companies were demanding exorbitant amounts of money to make connections and lay on supplies for them.

Around 1950, many farmers piped their own water into cowsheds because there was a great drive going on then to eradicate TB from cows, and having water on tap in the cowshed was one of the regulations. The government was paying a bonus of tuppence on every gallon of milk sold from attested TB-free herds. This was quite a lot of money, added on to the standard 1s 10d. As well as cows and milk being tested regularly for TB, there were also regulations about cowsheds and milking conditions. There had to be running water for washing down and for cooling the milk, and sheds had to have all walls and partitions made of concrete or other washable material, not wood. There also had to be a specified window area to let in adequate light. Sheds and milking equipment were regularly inspected for cleanliness. Any cow showing positive to a TB test had to be killed. Cardiganshire, together with one county in England, was the first in Britain to become completely TB free.

I live in Powys now, roughly 15 miles beyond the northern boundary of Cardiganshire. One of our nearest neighbours, half a mile away, is a Cardi lady. All around us are these alien creatures, sheep. The farming family who own the sheep and the land, the Butlers, very kindly allow us and Tom Dog to walk anywhere we like in the fields and woods, which is a great joy to me. We may also take any fallen timber from round about and this has kept our fires going these past 17 years. However, Howard does take delight in telling us of some of his smart deals, and Patrick likes to pull his leg by saying he's sure he must have Cardi connections somewhere in his past.

Chapter 21

Party Pieces

MADGE was very good at parties. They ranged from the mad, chase-around-the-house sort, to the quiet, music and talking sit-abouts. And then there were the outdoor singing-round-the-bonfire parties. One thing they all had in common was lots of good things to eat and drink.

There was one particularly lively occasion that was arranged as a harvest thanksgiving and a thank-you to people from the village who had helped us out with all kinds of jobs during the year. Some other friends from the village came to make up a good mix. They enjoyed some hearty chase-about games like Murder and Charades, and some quieter, pencil and paper stuff. Gareth from the Pentre astonished everyone with some rather good conjuring tricks. As it got near time to go home, they all started singing. And then Dai said he was going to sing *Bless This House*. He did this very solemnly, standing up with his eyes closed. After the third verse he lost track of the words and bending down said, 'What comes next, Austin?' 'Chimneys,' replied Austin. 'No. The chimneys are all right, bach. It's the roof I want.' And he picked up the thread again and finished in fine style.

Sometimes when Win and Charles were with us, we would have fun persuading Cecil to sing one of his old music hall songs and Win to recite *The Highwayman*. She did this very well but also added rather hiccuppy clip-clop noises for the horse's feet when the highwayman came riding. We all joined in, and Charles added funny comments. She rarely got through to the end for we all collapsed in laughter. One time there was a little girl with us. The very one who announced earlier that Sheila was upset she couldn't have a baby. This child kept nagging at Charles to do his party piece. He made various excuses about not having it with him, finally responding to, 'Where is it then?' by saying it was not well and was upstairs in bed. To which Win tartly replied that he'd better get it out before she went up there.

As well as music records, Stanley brought us records of Bernard Miles's country monologues. We played these over and over and repeated jokes and sentences from them like catchwords. There are still key words or situations that will prompt one of these stories or phrases from me and they have passed on to my daughter who knows them as well.

We did go to the cinema sometimes. A travelling one started coming round in 1948, I think it was, and it worked very well. I don't remember there being problems with breakdowns but the film did have to stop once or twice during the performance because the man in charge had to change the reels. The whole thing was run on a petrol-driven generator which the man brought with the rest of his equipment in a van. The generator was set up outside the church hall and the projector was connected and set up at the back inside. The screen was probably like a home projection screen, only bigger. Performances were always very well attended, for this was one of the most sophisticated forms of entertainment available. I was walked home from one of these performances one dark night by a young man, causing quite a lot of surprise among other friends.

'Well I never did!' a voice exclaimed as we passed by.

'Oh, you should try sometime,' I suggested.

My husband tells me that in his village the young lads used to go round to the generator while the film was showing and turn the fuel off. This caused everything to run down and stopped the performance. The wretched man had to go out and fiddle in the dark to start it up again. You can imagine the racket inside while the audience waited for the film to come back. We didn't have that kind of hooliganism in our village!

There were proper picture houses in the towns, but people from farther away could only visit these occasionally, and then only for an afternoon showing because they had to get there and back by bus. Few people had cars, certainly not the young ones who needed them. One or two had motorbikes, so could take a friend – if they were allowed out that late and had money and petrol coupons.

Cars, money and unrationed petrol must have all come at much the same time because there was a sudden rush of cars on the roads, probably starting in around 1950. Many more holiday-makers were coming to the area and local people were chasing off all over the place in their own cars. The effect was particularly noticeable in the lives of women, where a quiet emancipation went on. Well, quiet on the surface. There were likely to have been revolutionary ructions going on behind some doors. Ladies who had been relatively housebound and dependent on buses or husbands to take them about were now able to drive off to town whenever they felt like it, having hair dos, shopping, and getting jobs.

Local wages were better too, especially as young men were becoming more aware of their rights and their ability to stand up for them. One of the Pentre Arms boys, Morlais Jones, took up the cases of some of these men who had come to farms from children's homes years ago and had never been paid proper wages. He saw to it that they got compensated. One man who had come to a farm in the district in this way and later became very active himself on local councils, told me of his hard life of long hours and heavy work as a young boy. Getting his first year's wages – I think it was about £3, he went off to Cardigan for a day at the fair. He had to pay his bus fare, bought himself a suit to wear, had a beer and something to eat, then spent the rest of his money on rides at the fair.

'You spent all of it?' I found this quite astonishing.

'Yes. I was out to enjoy myself.'

'Oh well,' I said, 'easy come, easy go.'

It took him a long time to see the joke.

He maintained the fine walled garden belonging to Pigeonsford Mansion. When he had to give it up, it declined into a sad state of wilderness and disrepair for several years before being taken over by a couple of newcomers, David and Hilary Pritchard, who have made it good as new, or even better.

The annual fair was a great occasion, of course. It came around in the autumn, stopping a few days at all the little towns with its roundabouts, dodgems, swing-boats and all the supporting sideshows. I remember one lovely November day at the Cardigan fair, but as always we had to leave in the early evening to get the last bus home. I was with the young man who walked me home from the pictures. He bought a motorbike later and told me, 'You can stay all night at the next fair if you want to.'

We had quite a lot of live concerts and plays to go to in the village, and in neighbouring villages. Plenty of people could sing and recite, and performances were very good. A new vicar did an act one year. He had been a padre in the RAF. He talked about some places and incidents from the war and he told a couple of risqué service jokes. None of your really coarse, barrack-room stuff, but he was definitely not approved of and stayed only a few months until someone more suitable was found.

The Women's Institute (WI) provided a lot of entertainment. There were the usual interesting and helpful talks and demonstrations for members, but we also put on plays and other concerts and had country dancing evenings. I wasn't terribly keen to join the WI but went initially to keep Madge company. Madge hadn't thought it was her sort of thing either to start off with, but she had to do something active in the community to use her business and organisational skills. The village people were always open to new ideas, not at all clannish or small-minded. Though neither were they to be put upon. It was a very active community and we had lively discussions about local and wider issues.

There was one considerable battle with another organising lady. I think she must have been the Big Fish in the pond before Madge came. They locked jaws a few times before Madge ousted her in the battle of the United Nations. The lady was determined we should raise funds to give to the United Nations and Madge maintained this was against WI principles of independence from all political organisations. The lady concerned was always known to us afterwards as Mrs Uno. Madge became a much bigger fish in wider WI ponds as years went by, but what she was probably best remembered for was the holidays she organised. She and Tom Lewis, who ran the bus service, worked together planning routes and checking hotels. He drove the bus on many of these holidays and notably on the last one before she died. She told me how Tom drove her through the main shopping centre of Carlisle, or Fort William – I forget precisely which – parking his coach outside shops so she could walk in and buy presents she wanted to take home. When they came back, I collected her from the bus at Bala and drove her the remaining 80 or so miles by car to make a quicker, more comfortable journey. As we got ready to leave, people kept coming up to say goodbye and thank her for the 'best ever holiday.' It was very moving and I was happy for her to have such a good send off and to be able to enjoy it. She died a week later, comfortably at home with her family, and just 81 years old. Cecil had died 25 years earlier.

We put plays on at the village hall and entered WI competitions with some of them. One-act plays were generally rather feeble, especially those for women only, and we read through dozens of them trying to find something interesting and suitable. Luckily WI competition rules allowed us to include two men in the cast, which did make things easier. One year we came first in the inter-county competition with a dramatic tale about a hangman. It was called *Dark Brown*.

Some of the friends who came to stay at the farm were accomplished musicians, particularly as pianists. I owned a piano of sorts and it was got out of store and set up in the farm living room mainly so that we could enjoy performances by them. Les and

Prizewinning WI play, *Dark Brown*. Seated: Sheila and Mrs Evans-Green-Frills. Left to right: Little Mrs Jones-Who-Knows-Where-Her-Wheels-Are, Beryl Pentre Arms, Mrs Thomas Siop Canol, John Day, Ivor Parsons.

David were the main players and I acquired my great love of Chopin from those hours of lilting, bubbling, sometimes ferocious, music that was coaxed and beaten out of the old piano.

It got really disreputable and Madge started to niggle about getting rid of it. After all, it delighted us for such a short amount of time in the year and we did now have access to good quality music at all times from the electric wireless and radiogram. We talked about it and eventually agreed that I should invest some of my money in a new one.

So one day in mid-1951, I went off by bus and train to Swansea. Madge organised everything beforehand, giving me train and bus times, looking up music shops and giving me the names of shops and roads to help me on my way. I couldn't go wrong.

Well of course, I could always go wrong! And having set out to buy a piano – I came back having bought a Vespa scooter!

I walked out of the railway station and across the road, and there suddenly before me was this window full of beautiful bright motorbikes. They glowed, and I gazed entranced. That might have been the end of it, but Madge and I had been for a holiday in the Italian mountains three months earlier, and we had watched these tiny little bikes chugging past us with two huge bottoms overflowing the main seat and pillion.

Madge said, 'I wouldn't mind one of those! If they can carry two people that size up these mountains, one could carry me up our hills.' We talked about it semi-seriously, and I did think one might be very useful for me for getting about. So when they confronted me that day, I was half prepared to fall. I went in and did the deal and then went home to break the news to Madge and Cecil, and to wait for it to arrive. It didn't quite take the place of a piano, but I did get a lot of use and pleasure from it. I would have got even more if I'd learnt something about its workings and how to drive it properly. Instead of which I was forever burning its cork clutch out. Though that did have its interests as well. She and I did a lot of our travelling on kindly lorries. And there were the young men who towed me that time. Once, in a little village in Devon, a chemist cut slices off a variety of his bottle corks to fit in the clutch plates and put us back in running order. I spent part of that night helping a man to pack up bundles of daffodils for the spring flower market.

When I passed my test, I came out of the Pentre Arms some time later and found that a friend had loaded the Vespa with flowers in celebration. He then gallantly came for a ride on the back of it with me, the only male ever to do so. Tom, his name was. One supercilious fellow, when offered a lift, actually said he couldn't be seen out riding on the back of a lady's motorcycle! I was very nervous with my first passenger. He made a considerable difference to the balance. As we came down one steep hill and round a sharp bend at the bottom, I slowed down so much that we lost balance entirely and tipped gently over into the ditch and sat there laughing like idiots. It was anyway a very strangely balanced machine with the great weight of the engine on one side at the back. When I had a suitcase packed on the back as well, the thing used to rear up off the road and go round corners on its back wheel.

Madge did some outdoor parties with bonfires and sing-songs when there was a biggish crowd staying. And during the summers of 1952 and 1953 there started to be village beach parties with big bonfires.

There was a great street party in the village for the Coronation, with tea and games for the children, floodlights and fairy lights at night and dances in the garage.

When I got married a few weeks later, we didn't have a party at all. This was entirely my fault. I just wanted to get the business done as soon as possible and with as little fuss as possible. All I was doing was what everyone did at about my age and signing a bit of paper to register the fact. I had determined I wanted the date to be the ninth day of the ninth month. As this was in the middle of harvest, I didn't see why Madge should have to be bothered with any extra work on my account, though of course she would have loved it as she always did and would have made it a tremendous occasion. I didn't even expect a cake, though she did make a splendid one, intricately iced in between dashes to get the harvest in. And liberally spiced with 'sod its', she said.

One other thing I was trying to do was see if I could fool the village and actually do this thing under their noses without anyone knowing. And against all the odds and the underground intelligence system, I actually achieved it. I don't think anyone minded. As they got the news, they had a lovely time going round saying to others, 'Have you heard?' They seemed to get at least as much fun out of having been fooled as they would have done by talking about it beforehand. I did tell William Henry on the day so he wouldn't feel left out and could say, yes, he knew. Or, indeed, could do the telling.

And so I began another, very different, meander.

Llangrannog Carnival Queen, Eileen Price, with attendants, outside the Ship Inn.

Annibendod

ANNIE Lloyd gave us the word annibendod. It approximates to the English odds-and-ends. We liked it so much we used it as a name for one of our cows. This chapter is a few left over odds-and ends of thoughts.

And following on from the end of the last chapter, I have to mention an even more remarkable hoodwinking of the village than my wedding. This was Dot's baby. There she was one day walking down the street looking her usual elegant self, next day she had a baby daughter. And not a soul had recognised it was on the way. True the village social structure was beginning to crumble a bit by then, but there were still plenty of the old people left, schooled in the country lore of watching and listening and gossiping, and of putting two and two together and making five.

I suppose it was a little way into the 1960s when changes in the village began to be really noticeable. More and more of the old people were dying and more and more young people were moving away because new technology took local jobs and offered more interesting and exciting prospects elsewhere. For many years, the majority of houses in the village were empty most of the year. They were either owned by locals and rented out to holiday-makers, or they were holiday homes visited just once in a while by their owners. By 1970ish, the days had gone when I walked down the street knowing and chatting to everybody. Before then, summer visitors had been almost as well-known as locals, for they came, the same ones, year after year and were looked forward to as returning friends, almost honorary villagers. In recent years a more stable population has become established again. Retired people, home-workers, commuters, have made a new sort of village community, which seems to work well in its way, but I'm glad I'm not there to have it fighting with my memories.

There's a smart petrol station and emporium up at Brynhoffnant too, where for so many years was the old ramshackle garage and its petrol pumps. That's where we took our accumulators to be charged up so we could listen to the wireless, and our wellies to have holes patched. Everybody took their old motor cars there for Glyn to work his miracles of repair on. That was in the days when engines could be taken apart and almost everything was replaceable or mendable. Cars went on running forever, and in almost any emergency a judiciously placed hairpin, silk stocking or elastic band would get you home.

A few yards away another tin shed housed the local branch of the farmers' co-op. Here could be bought all the usual animal foods, fertilisers, chemicals, brushes and brooms, buckets, seeds, wellies, boots and clogs. Many people were still wearing clogs. They were very strong. The thick wooden soles were immune to water and carried the feet above the average amount of mud common to farmyards and the rough earth lanes. Leather uppers were attached to a ledge in the top of the sole and you could have an iron strip running round over this, plus iron toe-caps and iron strips under the sole and heel. All of these parts hung from the corrugated iron ceiling.

Another local commodity was cwlm or cwlwm. I don't know if it was used anywhere else than in the south and west counties of Wales, for it was a by-product of anthracite, and South Wales was the only place in Britain where this was mined. The fine anthracite dust was mixed with clay and water to make a strange but very effective fuel. It is, or was (what has become of it?), the fuel used in farmhouse open grates. I don't really know how it was manufactured in our day. I suppose men with shovels mixed it and turned it. Or maybe something like a cement mixer? But twice in old histories I have read that the women did the job centuries ago, putting together the dust and clay and stamping it around while adding whatever amount of water was required to get the consistency right. The books quoted warnings to coach travellers passing through small towns – Cardigan was once named – regarding this custom. Gentlemen were told they should draw the blinds across their carriage windows and on no account allow ladies to look down side streets they passed for there would be rough women with their skirts tied up round their waists, stamping about with bare legs and feet in this black labour.

The curious substance was packed into the grate on top of a shallow layer of hot embers until you had a solid block of slightly wet, black clay pudding with every semblance of being completely cold and dead. The next step was to push the poker down through it from top to bottom, making three air-holes. In the morning it would look exactly the same, possibly a little drier and deader. You then wiggled the poker about to enlarge the chimneys, and in an amazingly short time the whole thing would be glowing red-hot. Alternatively it could be quite out, but this was rare and a sign of poor management. I rather think we used to dribble paraffin down the hole when it did happen. Then a match and it soon burned up.

The big changes that have taken place in farming all have to do with technology and mechanisation. The first was the change over from horse to tractor power, but though it was in those years after the war that tractor use became virtually universal on farms, there had been steam engines doing some of the same kind of work since the 19th century. They were already well established in industry by the 18th century and were adapted for farm use during the 19th century. The huge heavy machines were originally pulled to their place of work by teams of horses, but very soon, before the end of the century, they became traction engines, moving along under their own power. They still worked as stationary machines, once in position. They were used to drive balers and threshing drums. Farmers brought their corn to the site to be threshed by the threshing drum and then might have the straw baled.

These traction steam engines also provided power to pull ploughs. They were set up, one at each end of a field, and pulled a plough up and down between them on a cable. As there were six or eight furrows being drawn out on each trip, this made for very fast ploughing. A man sat on the plough to steer a straight furrow. At the end of each row of turned furrows, both the steam engines and the plough moved one block along and started again. There were two sets of shares on the plough, pointing in opposite directions. One set ploughed up the field, the other down.

A small point of interest connected to steam engines was that, in their time, numerous ponds were dug all over the countryside to provide water for the running of them.

It was during World War One that the lighter, more manoeuvrable petrol-driven tractors started to be made. As with so many technological advances, they developed from wartime requirements, but in the years of farming depression between the wars, no ordinary small farmer could afford to own one.

All this changed during the 1940s. Farmers were helped and bullied, and actually paid, to produce more and more food to ensure that the people of the British Isles could survive without imported food. Tractors and machinery were hired out by the government and farmers got used to working with them. Then with their larger incomes they began to buy their own. Very soon they were producing enough to feed the nation on much more than the meagre wartime rations – only to find that it was no longer wanted. People were looking for more exotic tastes from overseas.

One exotic taste from his old life that Cecil greatly missed was his smelly cheeses. I have no idea what fancy cheeses were available during the war. It would seem that people accustomed to such things could still get them if they had the right contacts. Cecil didn't see why Mrs Thomas, who had our ration books, shouldn't be able to obtain some and he kept pestering her until one day she did produce a section of blue cheese. I don't know what breed it was. Stilton probably. Whether or not it had to count as part of the regulation 2oz per person per week, or if it came as an extra, I don't know. Anyway, he brought it home and set it triumphantly on the table and he and Madge enjoyed it. I think I had not got into serious cheeses at that stage. Another piece appeared the next week and another after that. Then round about that time, Cecil came back with a lump that must have been half the entire cheese.

'Funny thing,' he said, 'nobody else will buy it. Thosses Mimas (as we called her) asked me to take it away as people are complaining of the smell.' Personally I wasn't surprised.

It was put on the table for lunch that day. I can still see the expression on Henry's face when he was pressed to try some. It was certainly a fine ripe cheese. It sat there glistening and humming to itself. I gazed at it, wondering how anyone could enjoy eating something that smelt like that. With all that mouldy blue and slimy white, it reminded me of the damp wallpaper sliding off the walls. As I stared, bits of it seemed to be changing places with other bits. A small lump shook itself loose and fell in a writhing heap on the edge of the plate.

'It's alive,' I breathed.

'Ch-rist!' said Madge, in a voice even more outraged than I had ever heard her when Cecil swung the car too fast round a sharp bend. 'Take it out, Cecil.'

'It's all right,' he said mildly. 'Only wants a few bits cut away.'

'Get it off my table.' When Madge spoke in that white-hot ferocious voice, not even Cecil hesitated.

I fear it probably ended up in the pigs' troughs – though I could well imagine Cecil going down to the sty and sitting on the wall sharing it with them.

When I was a small child, there were two jobs I used to be given, both of them largely obsolete now, which I loathed. One was shelling peas, the other was scraping the scales off herrings. I'm sure it was herrings, though they are so smooth today I often wonder if I can be remembering rightly. I can't imagine what else they'd be if not herrings. It was

Traction engine running threshing drum, Essex, about 1912. Little Wakering church tower is in the background.

a horrible, slimy business. The scales were very difficult to get off and when they did come off they stuck on my hands and arms and under my fingernails. The problem with peas was quite different. In those years the pods always had maggots in them and I was utterly terrified of maggots. I had seen them seething on dead creatures in the countryside and I imagined them getting inside me. Almost every pod contained two or three and even some of the peas themselves had been burrowed into. The good peas had to be picked out and the best of the others could have the bit with the maggot in dug away with a thumbnail. What dreary, unexciting lives kids live these days!

I properly came to terms with maggots the day Madge gave a half scream and a shudder and passed one of the big juicy hams down to me from the ceiling hooks.

'Can you do something with that! The flies have got at it.'

I did not immediately equate this statement with maggots. Then I saw the little white wiggly pieces filling the space round the hipbone.

'Yes,' I said dismissively. 'There's only a few maggots round the bone. It'll be fine.'

I sat on the grass outside the house with a knife and a cloth and a bowl of water, and I scraped and washed and squashed until it was mostly only ham left.

'There you are – good as new,' I said, using the term applied locally to absolutely anything being sold second-hand. 'If there's any left they'll come out in the boil.' I may well have added, 'thas only a few tadpoles/maggots. We relies on them to help us out with the meat ration' – adapting Bernard Miles's countryman comments when told his well-water was undrinkable and full of 'microphobes and minute hanimal horganisations.'

There is one very important member of Madge's 'family' I haven't mentioned. This is Timmy. He features in some of the old photographs. He didn't properly join until after I left the farm and he had moved on to New Zealand before I went back. Having started coming up to help with odd jobs when he was about 10, he became much loved and useful about the place. He went to Agricultural College and later took over William Henry's job when he became ill. He had a game he used to play with the two corgis similar to Charles's with the chickens. He took them down to the barn, said 'Round the other side!' and when they had pounded away, he came out the same side he'd gone in. He still keeps in touch with Felicité. She still lives on, but detached from, the farm, in the house Madge had built, nicknamed The Dower (House).

Now only Beryl, of the Pentre Arms' Joneses, remains in the village. She and her sister, AnneJane, and their mother, long ago moved from the pub. One brother is alive of the original four, but he, like Morlais (meaning voice of the sea) and Gareth, lived most of his adult life away from the village. Bunty and Dot, old friends I had fun and happy times with, were still both there until recently when Bunty went to live with her daughter. Teifion is still there, who was at the fair that day with Wyndham and me. I met him recently, first time for 30 or 40 years and he gave me a kiss. 'It's a long time since we did that,' I said. Dear Morwen, William Henry's wife, is still there too. Sally Cefn Cwrt and young Geoff who worked for her and Jim, they're still about, but they were away from the village a bit and Geoff was too young for me to get to know well. What a lot of memories we hold between us of a completely different era. Where will they go when we die? Are those lives all out there still, in the wind and sea and cliffs? Available to see like a great film if someone would invent the special spectacles for viewers to put on? When Gareth died, I wrote to Beryl that if there was a heaven and a God, I reckoned all His most special angels would be Joneses and there'd be a huge picture of Llangrannog hanging on the Throne Room wall.

There certainly is in my heaven.

Index

Aberporth 29

Aberystwyth 171

Assize Court 149

Austin 143, 177

Austria 172

Bardsey Lighthouse 17

Barefoot, Sheila 28, 71

Bloody Sunday 143

Bridgend 43, 95, 133, 146

Brynhoffnant 26, 68, 145, 185

Burford 148

Burton, Jimmy 6, 83

Cambridge Low Temperature Research Station 44

Cardi 40-1, 169-71, 173, 175-6

Cardigan Bay 69, 164

Carmarthen 16, 25, 68, 79, 171-2

Carreg Bica 28, 30

Cefn Cwrt Farm 29

Ceredigion 56, 134, 170

Civil Service 57, 129

Cochen 40, 153

Collins Piddle 46

Commercial Hotel 171

Coronation 182

Cranes Ltd 11, 14

Cranogwen 6, 21

E27N Fordson Major 50, 54

Easter 78-9, 83, 144

Elsan 35-6, 130

Evans, Sir David Owen 63

Fellows, Eric 142

German POWs 67

Goose, Mrs 113-15, 117

Government Regulations 118

Hakesley, Joe 152

Henry, William 44, 80, 102, 113, 183, 189

Ianto 142

Ipswich High School 96

Jones, Griffy 6, 37, 72

Jones, Morlais 179

Lewis, Tom 180

Lister 89

Lloyd, Annie 49, 75, 111, 184

Lockley, Ronald 99

Machynlleth 75, 171

Maiden Lady 129

Master Mariner 21

Mermaid Rocks 157

Methodists 25

Miles, Bernard 178, 189

Milk Marketing Board 66, 174

Morwen 44, 189

Neale, Miss 96
New Quay 64, 69
New Year 30, 149
Newcastle Emlyn 57, 68, 172
Owen, Mrs 101
Pen Rhyp 158-9
Penbryn 32, 83, 130, 155-6, 159
Pentre Arms 44, 141, 145, 159, 179, 181-2, 189
Phillips, Miss 40
Pigeonsford Mansion 179
Pontgarreg 8, 172
Probert, Geoffrey 29
Pryce-Davies, Mr 173
Ragwort 100
Sally Cefn Cwrt 189
Seymour, John 71
Ship Inn 101, 183
Siop Canol 26, 181
Skokholm Island 99
Standard Fordson 50
Stooks 166, 168

Swansea 181
Teifion 189
Thomas, Mr 26, 101
Thomas, Mrs 26, 101, 181, 187
Thresher, Johnnie 50, 54
Threshing 47-8, 50-5, 168, 186, 188
Timmy 189
Troednoeth, Sheila 28
Ty Bach 35
Uno, Mrs 180
Vespa 150, 181-2
Welfare Officer 146
Welsh Cob 173-4
Welsh Parlour 103
Wesleyan Chapel 26
Woolwich Arsenal 29
World War One 186
World War Two 10-11, 174
Wyndham 189
Wynford 173
Ynys 97-9, 157